SpringerBriefs in Molecular Science

For further volumes:
http://www.springer.com/series/8898

Genxi Li · Peng Miao

Electrochemical Analysis of Proteins and Cells

 Springer

Genxi Li
Department of Biochemistry
 and State Key Laboratory
 of Pharmaceutical Biotechnology
Nanjing University
Nanjing
People's Republic of China

Peng Miao
Suzhou Institute of Biomedical
 Engineering and Technology
Chinese Academy of Sciences
Suzhou
People's Republic of China

ISSN 2191-5407   ISSN 2191-5415   (electronic)
ISBN 978-3-642-34251-6   ISBN 978-3-642-34252-3   (eBook)
DOI 10.1007/978-3-642-34252-3
Springer Heidelberg New York Dordrecht London

Library of Congress Control Number: 2012950863

Printed on acid-free paper

Springer is part of Springer Science+Business Media (www.springer.com)

Preface

The goal of this brief is to introduce modern electrochemical analytical techniques and the applications for the analysis of proteins and cells. So far, a large number of papers have been reported on the analysis of proteins and cells, which play important roles in all biological sciences, such as biochemistry, cell biology, molecular biology as well as biomedical fields like cancer research. An in-time summary is urgently needed. We therefore planned to publish a brief on this topic. The main emphasis is on the principles of electrochemical strategies and the practical utility of the detection systems. This brief offers an up-to-date, easy reading presentation of recent advances on the subject and is suitable to be used as a supplement for graduate-level courses in analytical chemistry, biochemistry, biophysics, biotechnology, biomedical engineering, etc. It may also help young scientists to get an overview on this topic.

Genxi Li
Peng Miao

Contents

Abbreviations

αDR	Anti human DR monoclonal antibody
AChE	Acetylcholinesterase
AIDS	Acquired immune deficiency syndrome
AuNPs	Gold nanoparticles
BSA	Bovine serum albumin
CEA	Carcinoembryonic antigen
CMEs	Chemically modified electrodes
CNTs	Carbon nanotubes
Con A	Concanavalin globulin A
cyt c	Cytochrome c
DPV	Differential pulse voltammetry
EGFR	Epidermal growth factor receptor
ELISA	Enzyme linked immunosorbent assay
GCE	Glassy carbon electrode
HRP	Horseradish peroxidase
hTf	Human serum transferrin
IDO	Indoleamine 2,3-dioxygenase
ITO	Indium tin oxide
JUG$_{thio}$	5-Hydroxy-3-hexanedithiol-1,4-naphthoquinone
MUC1	Human sticky protein-1
MWCNTs	Multiwalled carbon nanotubes
NADH	Dihydronicotinamide adenine dinucleotide
OP	Organophosphorus
PDGF	Platelet-derived growth factor
PEG	Polyethylene glycol
PFV	Protein film voltammetry
PGE	Pyrolytic graphite electrode
SAM	Self-assembly monolayer
SELEX	Systematic evolution of ligand by exponential enrichment
ssDNA	Single strand DNA
SWCNTs	Single-walled carbon nanotubes

Chapter 1
Introduction

Abstract The analysis of proteins and cells is of great importance in all areas of life sciences, so it has received more and more research interests of the scientists from various fields. Meanwhile, a number of different techniques have been exploited for the analysis of proteins and cells including mass spectroscopy, electrochemistry, ultraviolet (UV) spectrophotometry, flow cytometry and so on. Among these techniques, electrochemistry is a powerful means to study the electron transfer process in biological systems. With simple, rapid, sensitive and easily operated features, electrochemical technique has been applied in the related research fields for many years. In the meantime, some correlative techniques such as surface modification, molecular self-assembly, molecular recognition and nanotechnology have been developed rapidly, which have further broadened the applications of electrochemical technique in the field of the analysis of proteins and cells.

Keywords Electrochemical analysis • Protein activity • Cell detection • Analytical chemistry • Nanotechnology • Fundamental of electrochemistry

Along with the explosive development of human society and technology, more and more reliable, efficient and conveniently operated assays of various substances are urgently needed. In the fields of biology and medicine, analysis of proteins and cells has attracted particular interests.

Proteins are essential components of cells and organisms. Protein is consisted of polypeptides, which are linear polymer chains of amino acids. They usually fold into a globular or fibrous form. Enzymes are a major class of proteins catalyzing metabolic reactions, while some non-enzyme proteins have structural or mechanical functions. For example, actin and myosin in muscles are indispensable proteins in motor system.

Proteins participate in almost all processes inside and outside cells, such as cell signaling, cell cycle, immune responses and so on. For instance, mitogen-activated protein kinases pathway is an important cell signaling pathway, where three protein kinases are involved [1]. Through consequent phosphorylation by the kinases, the upstream signals can be transmitted to the downstream response molecules [2, 3]. The pathway plays a critical role in the regulation of gene expression and cytoplasmic functional activities. Abnormal protein expression levels may disturb cell functions and lead to a lot of diseases [4]. Therefore, the analysis of proteins is of significant importance in both scientific research and medical diagnosis of diseases.

The cell is the so-called building block of life, so it is highly necessary and greatly important to analyze the functional components inside cells and assay how cells work in different stages. The most direct way for the analysis of cells is to observe them under microscopes, including optical microscope, transmission electron microscope, scanning electron microscope, fluorescence microscope and confocal microscope. There are also some other techniques for the analysis of cells, such as immunohistochemistry, computational genomics, transfection, flow cytometry, immunoprecipitation and so on.

However, the applications of many traditional analytical techniques are usually restricted due to time-consuming process, complicated operation, low sensitivity and/or high cost. Electrochemistry may overcome many disadvantages of those traditional methodologies. Besides the dramatically improved sensitivity and selectivity, electrochemical method may have some inherent benefits for biological research, since many biological events like signal transduction, energy conversion and metabolism concern with the changes of current or potential *in vivo*. So, electrochemical analysis can intuitively reflect the oxidation or reduction of specific analytes such as ions, oligonucleotides and proteins [5–7], and the measurements of electrochemical parameters like potential, current or charge can be used for many more analytical purposes [8–10]. Therefore, as an interdisciplinary field from electrochemistry, analytical chemistry, biochemistry, biophysics, biomedicine, etc., electrochemical analysis of proteins and cells has attracted broad and extensive research interests.

The origin of biological analysis with electrochemical technique can be traced to 1962, when Clark and Lyons proposed a novel design combining an enzyme (glucose oxidase) and an oxygen electrode for the electrochemical detection of glucose [11]. Since then, more and more sophisticated electrochemical systems have been fabricated for the biological analysis based on the fundamental of electrochemistry. Nevertheless, since the electron transfer rates between proteins and electrode surfaces are usually prohibitively slow due to the busying of the electroactive prosthetic groups in the electrically insulated peptide backbones and the unfavorable orientations and adsorptive denaturations of proteins, electrochemical analysis with proteins is often difficult to be performed, not to mention the assay of the biological macromolecules themselves [12, 13].

In recent years, with the rapid development of protein electrochemistry together with the related technologies such as surface modification, molecular recognition and assembly, nanotechnology, etc., electrochemistry is gradually moving beyond some disadvantages in the study of proteins. Meanwhile, many of the molecular bases of biological processes and/or diseases are disclosed. Therefore, more and more analysis of proteins, and even cells, with electrochemical technique has been conducted [14–21]. In this Springer brief, the authors will give an overview to this interdisciplinary field based on the results of some typical studies carried out by the authors and some colleagues.

References

1. Seger R, Krebs EG (1995) The MAPK signaling cascade. FASEB J 9(9):726–735
2. Manning G, Whyte DB, Martinez R, Hunter T, Sudarsanam S (2002) The protein kinase complement of the human genome. Science 298(5600):1912–1934
3. Sebolt-Leopold JS, English JM (2006) Mechanisms of drug inhibition of signalling molecules. Nature 441(7092):457–462
4. Antic D, Keene JD (1997) Embryonic lethal abnormal visual RNA-binding proteins involved in growth, differentiation, and posttranscriptional gene expression. Am J Hum Genet 61(2):273–278
5. Davis J, Moorcroft MJ, Wilkins SJ, Compton RG, Cardosi MF (2000) Electrochemical detection of nitrate and nitrite at a copper modified electrode. Analyst 125(4):737–741
6. Fan CH, Zhuang Y, Li GX, Zhu JQ, Zhu DX (2000) Direct electrochemistry and enhanced catalytic activity for hemoglobin in a sodium montmorillonite film. Electroanal 12(14):1156–1158
7. Wang QX, Zheng MX, Shi JL, Gao F, Gao F (2011) Electrochemical oxidation of native double-stranded DNA on a graphene-modified glassy carbon electrode. Electroanal 23(4):915–920
8. Dilimon VS, Fonder G, Delhalle J, Mekhalif Z (2011) Self-assembled mono layer formation on copper: a real time electrochemical impedance study. J Phys Chem C 115(37):18202–18207
9. Fan CH, Liu XJ, Pang JT, Li GX, Scheer H (2004) Highly sensitive voltammetric biosensor for nitric oxide based on its high affinity with hemoglobin. Anal Chim Acta 523:225–228
10. Zhou L, Glennon JD, Luong JHT, Reen FJ, O'Gara F, McSweeney C, McGlacken GP (2011) Detection of the pseudomonas quinolone signal (PQS) by cyclic voltammetry and amperometry using a boron doped diamond electrode. Chem Commun 47(37):10347–10349
11. Clark LC, Lyons C (1962) Electrode systems for continuous monitoring in cardiovascular surgery. Ann NY Acad Sci 102(1):29–45
12. Li GX (2006) Protein-based voltammetric sensors. In encyclopedia of sensors. American Scientific Publishers, United States of America
13. Li GX (2010) Heme protein-based electrochemical biosensors. In: Handbook of porphyrin science. World Scientific Publishing, Singapore (Heme proteins)
14. Shen M, Yang M, Li H, Liang ZQ, Li GX (2012) A novel electrochemical approach for nuclear factor kappa B detection based on triplex DNA and gold nanoparticles. Electrochim Acta 60:309–313
15. Yang NN, Cao Y, Han P, Zhu XJ, Sun LZ, Li GX (2012) Tools for investigation of the RNA endonuclease activity of mammalian argonaute2 protein. Anal Chem 84(5):2492–2497
16. Ahmed MU, Hossain MM, Tamiya E (2008) Electrochemical biosensors for medical and food applications. Electroanal 20(6):616–626
17. Masarik M, Stobiecka A, Kizek R, Jelen F, Pechan Z, Hoyer W, Jovin TM, Subramaniam V, Palecek E (2004) Sensitive electrochemical detection of native and aggregated alpha-synuclein protein involved in Parkinson's disease. Electroanal 16(13–14):1172–1181
18. Pividori MI, Lermo A, Bonanni A, Alegret S, del Valle M (2009) Electrochemical immunosensor for the diagnosis of celiac disease. Anal Biochem 388(2):229–234
19. Li GX (2007) Protein-based biosensors using nanomaterials. In: Nanotechnologies for life sciences. Wiley-VCH, Weinheim (Nanomaterials for biosensors)
20. Zhao J, Zhu L, Li XX, Bo B, Shu YQ, Li GX (2012) An electrochemical method to assay the reversal effect on multi-drug resistance in tumor cells. Electrochem Commun 23:56–58
21. Yang JH, Zhao J, Xiao H, Zhang DM, Li GX (2011) Study of hemoglobin and human serum albumin glycation with electrochemical techniques. Electroanal 23(2):463–468

Chapter 2
Theoretical Background of Electrochemical Analysis

Abstract An electrode is a conductor or semiconductor, which directly contacts the electrolyte solution. In an electrochemical system, the input and output are both realized through an electrode. The substrate materials of the commonly used electrodes include noble metals (platinum, gold, silver, etc.), mercury, various kinds of carbon materials and semiconductor materials. Since the electron transfer rate between proteins and electrode surfaces is usually prohibitively slow, due to the burying of the electroactive prosthetic groups of most proteins in the electrically insulated peptide backbones and adsorptive denaturation of proteins on electrode surface, chemically modified electrodes (CMEs) are developed to facilitate the electrochemical analysis of the biomacromolecules and cells. Meanwhile, different electrochemical techniques are employed to meet the requirements of different bioassays.

Keywords Electrode substrates • Chemically modified electrodes • Cyclic voltammetry • Differential pulse voltammetry • Electrochemical impedance spectroscopy • Chronocoulometry

2.1 Electrode Substrates

The basis of electrochemical analysis is the reaction on an electrode surface. Since the working electrode substrates can strongly influence the efficiency of the reactions, the properties of electrode substrates are of great significance for successful electrochemical analysis.

So far, a variety of electrode substrates have been exploited. Noble metals, mercury, carbon and semiconductor materials are commonly used electrode materials. Some composite materials with excellent characteristics have also been employed for electrode substrates, such as platinum-dispensed titanium, tantalum and niobium. Besides, electrode materials with small dimension (e.g., a nanometer level) may greatly enhance the electron transfer rate and thus have superiority in lots of applications. For example, nanocrystalline tin oxide electrodes have been constructed to study the electrochemistry of certain redox proteins [1, 2].

Since different functional groups on electrode substrate possess different physical and chemical properties as well as biocompatibility, a protein usually exhibits extremely different electrochemical performance on different electrode substrates. Therefore, choosing appropriate electrode substrate is crucial for successful electrochemical analysis. While the obtained experimental results can help researchers choose suitable substrate for specific applications, prediction of the utility of electrode substrate can also be helpful, which is mainly based on two factors. One is the redox behavior of the target analyte, and the other is the background current over the potential range applied in the measurements. Meanwhile, some other factors should also be considered, such as the electrical conductivity, mechanical properties, toxicity, cost, etc.

2.1.1 Metal Electrodes

Nobel metals such as platinum, gold and silver have been widely used as electrode substrates. Noble metal electrodes can offer very favorable electron transfer kinetics and a wide anodic potential range. The cathodic potential window of these electrodes is usually restricted due to the low hydrogen overvoltage. However, the formation of surface oxide or adsorbed hydrogen layers may lead to high background currents, which strongly affect the kinetics of the electrode reaction. To solve this problem, a pulse potential cycle should be performed before electrochemical experiments [3]. Here, we will briefly discuss several typical metal materials.

The chemical properties of gold electrode and platinum electrode are very stable. Meanwhile, pure gold or platinum materials are easily obtained, and the electrodes can be conveniently manufactured. Therefore, these electrodes become most popular metal electrodes. Silver is also good electrode substrate, which is usually used for the preparation of chemically modified electrodes (CMEs) in various electrochemical researches [4–8]. Moreover, some proteins such as cytochrome c (cyt c) may exhibit the capability of direct electron transfer on silver electrode, so silver substrate has also been directly used for protein analysis [9].

Besides noble metal electrodes, some other metals have also been employed as electrode substrates. For instance, copper electrode and nickel electrode have been constructed for the detection of carbohydrates or amino acids in alkaline media. Compared with platinum or gold electrodes, these two kinds of electrodes possess a stable response for carbohydrates at constant potentials [10]. In addition, alloy electrodes like platinum–ruthenium and nickel–titanium electrodes have also been reported, which are often used for the preparation of fuel cells, owing to their bifunctional catalytic mechanism [11].

2.1.2 Mercury Electrodes

Mercury is a classic electrode material. With high hydrogen overvoltage, it can extend the cathodic potential window. Meanwhile, mercury electrodes possess highly reproducible, renewable and smooth surface, which is very beneficial in electrochemical analysis. So, a variety of mercury electrodes including dropping mercury electrode, hanging mercury drop electrode and mercury film electrode have been developed. Among the mercury electrodes, dropping mercury electrode is the most commonly used one. The main advantage of dropping mercury electrode is that the electrode can be self-renewing, so it does not need to be cleaned or polished before each experiment. Moreover, each drop of mercury has an uncontaminated and uniform surface. Nevertheless, the toxicity of mercury and the limited anodic range have restricted the application of dropping mercury electrode and the other mercury electrodes in the analysis for biologic species. Therefore, the colleagues have exploited some related solid amalgam electrodes for the biologic analysis.

2.1.3 Carbon Electrodes

A family of carbon materials have been widely used as electrode substrates to make various electrodes. Due to the soft properties of carbon, these electrodes surface can be easily renewable for electron exchange. Carbon materials also have broad potential window, low background current, rich surface chemistry and comparative chemical inertness. The cost of carbon materials is also very low. Therefore, carbon electrodes are currently widely used, and a large number of research projects are even focused on the relationship between structure and reactivity of carbon electrodes. The commonly used carbon electrodes include pyrolytic graphite electrode (PGE), glassy carbon electrode (GCE), carbon paste electrode, carbon fiber electrode and electrodes composed of carbon composites. All these electrodes have the basic structure of a six-membered aromatic ring and sp^2 bonding. One difference between these carbon electrodes is the relative density of the edge and basal planes of the surface which affects the electrochemical reactivity at electrode surface. For example, an elegant cyclic voltammogram of cyt c can be obtained at an edge plane PGE [12] and GCE [13], while only a very small response can be obtained on a basal plane PGE.

2.2 Chemically Modified Electrodes

In many research works, proteins under investigation are usually immobilized onto the surface of an electrode. However, this immobilization procedure may denaturalize most proteins with the conformational change, which may affect the further analysis of the proteins. Therefore, bare electrodes are not ideal interfaces to obtain direct electrochemistry of most proteins; thus, CMEs are developed to improve the situation. CMEs emerged in 1973 when Lane and Hubbard modified various olefine compounds on clean platinum electrode through chemisorption, which significantly changed the electrochemical response of the electrode [14, 15]. Since then, CMEs have been developing rapidly to investigate the direct electrochemistry of proteins and the mechanisms of redox reactions. The fabrication of CMEs is to immobilize molecules with specific functions on the ordinary electrode surface by chemical or physical methods. Proteins deposited on the surface of the CMEs can then retain their biologic activities to some extent; thus, electrochemical performance of the electrode can be improved for the analysis of proteins. For example, thiol compounds can be covalently attached to the surface of gold electrode through the formation of metal–S (Au–S) bond [16]. This process has then introduced some new functional groups not only for the improvement of biocompatibility but also for the later electrostatic interaction. Therefore, biologic analysis by using the CMEs can be conducted.

2.2.1 Conducting Polymer–Modified Electrodes

Conducting polymers are organic polymers with metallic conductivity or semiconductors properties. Since the discovery of high conductivity in doped polyacetylene in 1977 [17], numerous well-characterized and ordered conducting polymers have been exploited; thus, many conducting polymer–modified electrodes have been prepared for the analysis of proteins and cells.

The preparation of conducting polymer–modified electrode is often realized via in situ electropolymerization from monomer solution. Moreover, the properties of polymers can be modulated by attaching different chemical groups to the monomers before polymerization. Therefore, these chemical groups may participate in molecular recognition or electrocatalytic reaction, which can help the polymers become efficient molecular interfaces between electrodes and solution. Meanwhile, conducting polymer nanowires have also been synthesized and applied in the fabrication of resistance sensors and molecular electronic devices for the analysis of proteins and cells [18].

There are many advantages of conducting polymer–modified electrodes. On the one hand, most proteins can well function in non-aqueous media with high activities [19]. On the other, conducting polymers usually contain electronic states that can be reversibly occupied and emptied with electrochemical techniques [20], which can facilitate the electrochemical measurements of analysis for proteins and cells.

2.2.2 Self-Assembly Monolayers

Self-assembly is a term to describe processes that a number of spatially disordered objects arrange themselves in an ordered pattern via local interactions. The interactions include ionic bond, covalent bond, metallic bond, as well as weak interactions (e.g., hydrogen bond, van der Waals force and π–π interaction). The self-assembly system is not freestanding. It needs solid support, so electrode is an excellent support which can further play more function roles. Figure 2.1 illustrates the formation a self-assembly monolayer (SAM) on a substrate surface.

Fig. 2.1 Schematic representation of the formation of a self-assembly monolayer at a substrate. Reprinted with the permission from Ref. [22]. Copyright 1995 American Chemical Society

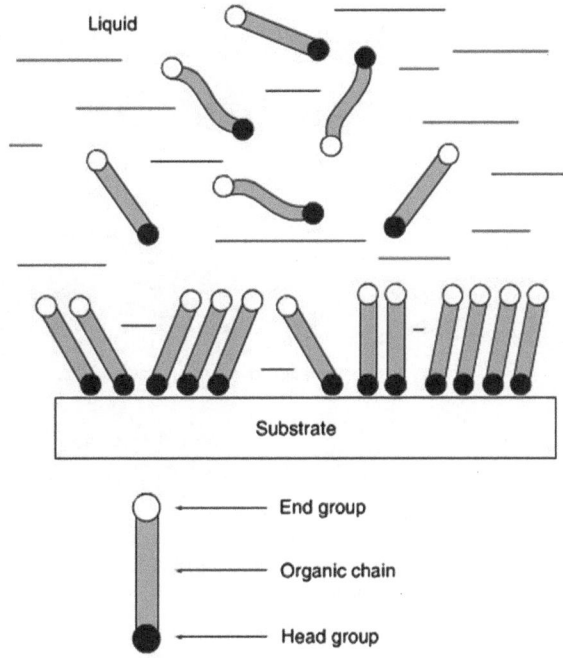

Self-assembly has been a widely studied surface modification technology, so SAM has been rapidly developed since the late 1980s and used in many scientific fields such as material science, molecular biology and medicine [21, 22]. Some properties like the injection across the interface between the monolayer and the electrode have been determined by molecular orientation and packing at the interface [23, 24]. Meanwhile, it has been known that the functions of the devices fabricated with SAMs usually depend on the deposition of the SAMs [25–27].

SAM has a lot of advantages. For instance, the formation of a SAM just requires a simple procedure, and the monolayer is chemically stable and biocompatible for electrochemical analysis. Moreover, proteins can be immobilized on the SAM-modified electrode with much more appropriate orientation compared with the adsorption on a bare electrode or in a polymer, overcoming the problem of denaturalization of proteins. Some electro-inactive reagents can also be used to help the formation of a SAM. Besides, a SAM can induce proteins to form an appropriate orientation, so the electron transfer rate between proteins and electrode can be largely accelerated accordingly [28].

2.2.3 Nanomaterial-Modified Electrodes

Nanomaterials possess at least one dimension sized from 1 to 100 nm [29]. They possess not only unique geometric, mechanical, electronic and chemical

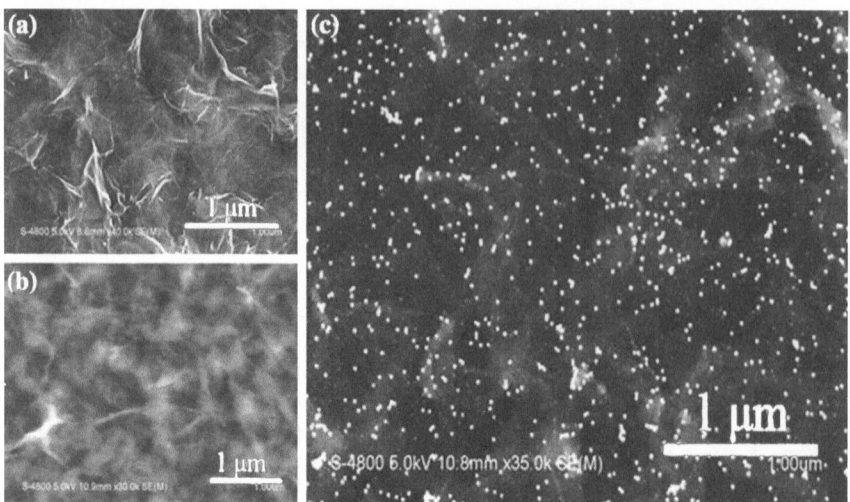

Fig. 2.2 Scanning electron microscope images of **a** graphene, **b** polythionine/graphene and **c** AuNPs/polythionine/grapheme film. Reprinted from Ref. [37], with kind permission from Springer Science + Business Media

properties, but also properties different from macroscopic materials, such as quantum effect, surface effect, small size effect, etc. These properties have greatly prompted a broad range of applications of nanomaterials in medicine, electronics, biomaterials, environmental science, energy production and biosensors [30–44].

Nanomaterial-modified electrodes have many advantages over the traditional material–modified electrodes. Firstly, nanomaterials offer huge specific surface area for the immobilization of more functional molecules on the electrodes (Fig. 2.2). Secondly, some semiconductor nanomaterials may act as promoters of electrochemical communications, accelerating the electron transfer rate between proteins and electrodes. Thirdly, some biocompatible nanomaterials can help proteins or cells maintain their activities on the electrode for a long period. Therefore, a variety of nanomaterials have been synthesized and characterized for the performance of electrochemical analysis for proteins and cells. The frequently used nanomaterials include metal nanomaterials, especially gold nanoparticles (AuNPs), metallic oxide/sulfide nanomaterials, carbon nanotubes (CNTs), especially multiwalled carbon nanotubes (MWCNTs) and graphene, etc. Moreover, nanocomposite materials, consisting of two or more types of nanomaterials, have also been exploited.

2.2.4 Mediator-Modified Electrodes

Redox reactions on the surface of an electrode can be accelerated by using a suitable electron transfer mediator [45, 46]. The function of the mediator is to facilitate the charge transfer between the electrode and the analyte. The reaction can be described as follows, in which M represents the mediator and A is the analyte:

$$M_{ox} + ne^- \rightarrow M_{red}$$

$$M_{red} + A_{ox} \rightarrow M_{ox} + A_{red}$$

Compared with bare electrodes, mediator-modified electrodes have the following advantages. Firstly, it can reduce the overpotential of the analyte and the possible interfering background current. Secondly, the response of current signal can be enhanced; thus, lower detection limit can be achieved. Thirdly, adsorption of the analytes and the products can be eliminated. Therefore, the sensitivity and selectivity of the analysis by using mediator-modified electrodes can be greatly improved.

Up to now, a lot of studies have been carried out for electrochemical analysis of various substances, including proteins, by using the electrodes modified with different kinds of mediators. For instance, sensitive and selective detections of dihydronicotinamide adenine dinucleotide (NADH) [47–49], ascorbic acid [50–53], cyt c [54–56] and hydrogen peroxide [57–60] have been achieved with the help of the corresponding mediators modified on the surfaces of PGE, GCE, gold electrode or platinum electrode.

2.2.5 Sol–Gel Technology

The sol–gel process is a wet-chemical technique, which has been widely used in materials science and ceramic engineering. This technique can fabricate an integrated three-dimensional network of materials like metal oxides from a colloidal solution via hydrolysis and polycondensation reactions [61]. Inorganic silica sol–gel materials are traditional sol–gel matrices, exhibiting excellent properties such as chemical inertness, high thermal stability and tunable porosity, etc. Nanomaterials have also been often used in the fabrication of complicated sol–gel three-dimensional networks. Since the formed structure of sol–gel matrix can maintain the native functional characteristics of the immobilized proteins [62, 63], a variety of studies have been conducted for the analysis of protein and cells based on sol–gel technology [64–67].

2.3 Electrochemical Cell

An electrochemical cell is used to generate voltage and current from chemical reactions or induce chemical reactions by the input of electrochemical signals. The most commonly used electrochemistry system is the three-electrode system consisted of working electrode, reference electrode and auxiliary electrode [68, 69]. Schematic illustration of the three-electrode system has been shown in Fig. 2.3. The

Fig. 2.3 Schematic diagram of an electrochemical three-electrode system:
a the reference electrode, **b** the working electrode, **c** the counter electrode, **d** the working solution, **e** constant temperature bath

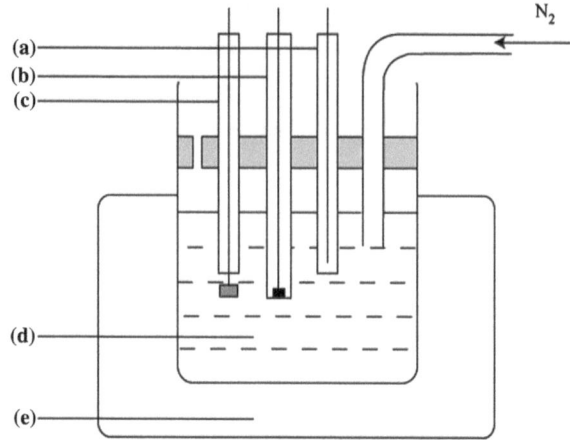

working electrode makes contact with the analyte. Its surface is the place where the reaction occurs. After the working electrode is applied with a certain potential, the transfer of electrons between electrode and analyte takes place. The current observed at the electrode will pass through the auxiliary electrode for balance. Inert conducting materials such as platinum and graphite with comparably large surface areas are usually used to make auxiliary electrode. The reference electrode has a known reduction potential, while no current passes through it. It only acts as a reference when measuring the working electrode potential. Silver–silver chloride and the saturated calomel reference electrodes are commonly adopted in the three-electrode system. To avoid the contamination of the sample solution, the reference electrode can be insulated from the sample reaction via an intermediate bridge.

2.4 Electrochemical Measuring Techniques

For electrochemical analysis of biologic substances, various electrochemical signals such as current, potential, charge and impedance have to be generated via biorecognition or biocatalysis processes, which are further related to the concentration of the analytes. So, the most commonly used electrochemical measuring techniques should be briefly discussed, including cyclic voltammetry, differential pulse voltammetry (DPV), electrochemical impedance spectroscopy and chronocoulometry [68, 69].

2.4.1 Cyclic Voltammetry

Cyclic voltammetry may provide the information of the thermodynamics of redox processes, adsorption processes and the kinetics of electron transfer reactions. It is the

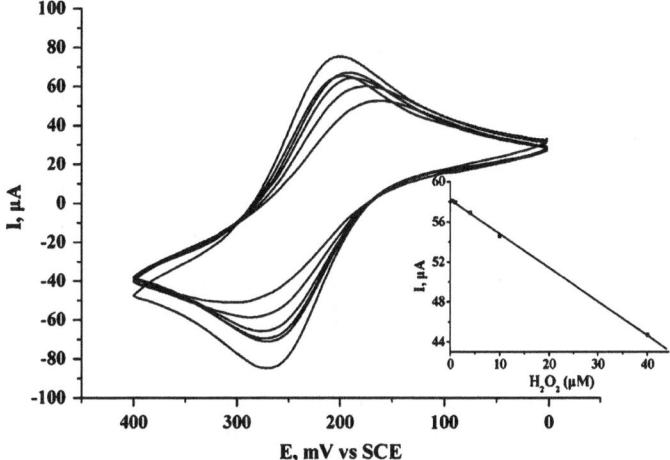

Fig. 2.4 Cyclic voltammograms of $K_4Fe(CN)_6/K_3Fe(CN)_6$ solution obtained at Au/cysteamine/ AuNPs electrodes upon the treatment by the AuNPs growth solution consisting of 0, 0.1, 1, 10, 100, 1,000 μM H_2O_2 (*from outer to inner*). Inset shows the linear relationship between the cathodic peak current and the concentration of H_2O_2. Reprinted with the permission from Ref. [38]. Copyright 2006 American Chemical Society

most widely used measuring technique in electrochemical analysis. In a typical cyclic voltammetry, the impulse potential is ramped linearly versus time and a pair of well-defined redox peaks are observed. Single or multiple cycles can be performed depending on the requirements of specific analysis. During cyclic voltammetric scanning, analytes or the redox centers of proteins will carry out certain electron communication with the electrode under various potentials and the currents may be proportional to the concentration of the analytes. Figure 2.4 shows a series of cyclic voltammograms of $K_4Fe(CN)_6/K_3Fe(CN)_6$ solution obtained at an Au/cysteamine/AuNPs electrode with the treatment by AuNPs growth solution containing different amount of H_2O_2. The peak currents are inversely proportional to the concentration of H_2O_2 [38], demonstrating cyclic voltammetry a fine quantitative analytical method.

2.4.2 Differential Pulse Voltammetry

DPV is a derivative of linear sweep voltammetry and staircase voltammetry, which is extremely useful to detect trace levels of organic and inorganic analytes. In this technique, there are a series of regular voltage pulses superimposed on the potential linear sweep or stair steps. Just before each potential change and late in the pulse life, the currents are recorded. The current difference is then plotted against the applied potential. In the differential pulse voltammogram, the height of the current peak can be directly proportional to the concentration of corresponding

Fig. 2.5 The basic
equivalent circuit of
electrochemical impedance
spectroscopy of an
electrolytic cell

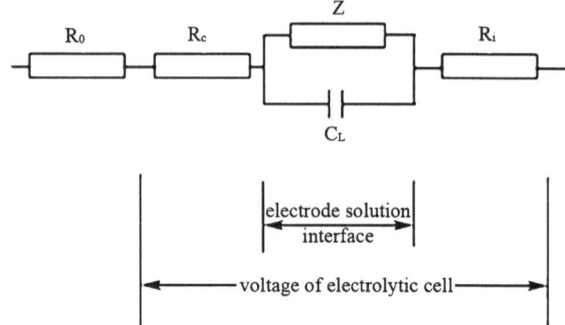

analytes. The peak potential varies with different analytes, which can also be used
to distinguish the detected species. DPV can not only help improve the sensitivity
of the detection and the resolution of the voltammogram, but also provide information about the chemical form of the analytes, such as oxidation and complexa-
tion status, which is very important for an analysis. Therefore, this technology has
also been widely used for the electrochemical analysis of proteins and cells.

2.4.3 Electrochemical Impedance Spectroscopy

Electrochemical impedance spectroscopy detects the dielectric properties of a
medium over a range of frequencies. The obtained data can be plotted in a Bode plot
or a Nyquist plot. The electrochemical process on an electrode can be simulated to
an equivalent circuit consisting of resistors and capacitors. Applied alternating voltage can generate alternating current from the electrochemical reaction on the electrode. The same alternating current can be produced if the voltage is applied on an
equivalent circuit. Therefore, the electrochemical behavior on the electrode is equivalent to a resistance, known as Faraday impedance (Z). Figure 2.5 describes the basic
equivalent circuit of the electrochemical impedance spectroscopy of an electrolytic
cell. C_L represents the capacitance of the electric double layer on the electrode, R_i
represents the internal resistance of the electrolytic cell, R_c represents the polarization
resistance of the electrode itself and R_0 is the resistance of the electrolytic cell circuit.
Usually, R_c and R_i are small, which can be neglected. Electrochemical impedance
spectroscopy can not only probe into the features of the surfaces of CMEs, but also
reveal the information about the reaction mechanism of an electrochemical process.

2.4.4 Chronocoulometry

Chronocoulometry is a technique measuring the relationship between charge
and time, involving stepping the potential of the working electrode from the

Fig. 2.6 Chronocoulometric curve and corresponding formula

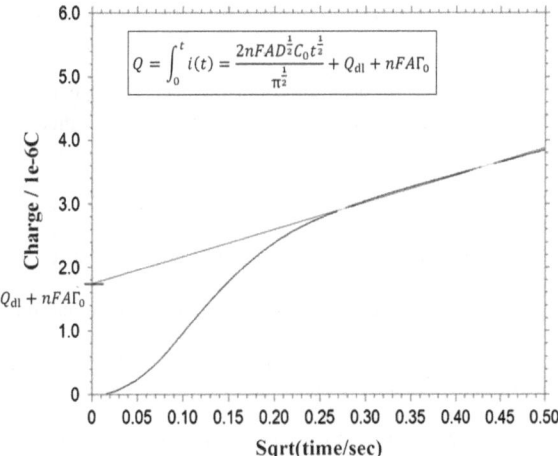

value at which no faradaic reaction occurs to the value at which the surface electroactive species concentration is zero. The adsorption phenomenon can also be characterized by chronocoulometry. In an electrochemical system, the function of the charge (Q) of the surface-confined redox species and the time (t) is described in the Cottrell expression as is shown in Fig. 2.6, in which n represents the number of electrons per molecule in the reduction, F is the Faraday constant (C/equiv), A is the electrode area (cm^2), D is the diffusion coefficient (cm^2/s), C_0 is the bulk concentration (mol/cm^2), Q_{dl} is the capacitive charge (C) and $nFA\Gamma_0$ is the charge from the reduction of redox marker on the electrode. Γ_0 represents the amount of redox marker on the electrode. When $t = 0$, the detected charge is the sum of the double layer charging and the surface excess terms. Under certain experimental condition, Q_{dl} is constant, so the quantitative determination of electroactive species adsorbed on the electrode can be achieved.

References

1. Astuti Y, Topoglidis E, Gilardi G, Durrant JR (2004) Cyclic voltammetry and voltabsorptometry studies of redox proteins immobilised on nanocrystalline tin dioxide electrodes. Bioelectrochemistry 63(1–2):55–59
2. Topoglidis E, Astuti Y, Duriaux F, Gratzel M, Durrant JR (2003) Direct electrochemistry and nitric oxide interaction of heme proteins adsorbed on nanocrystalline tin oxide electrodes. Langmuir 19(17):6894–6900
3. Johnson DC, Lacourse WR (1990) Liquid-chromatography with pulsed electrochemical detection at gold and platinum-electrodes. Anal Chem 62(10):A589–A597
4. Bas B, Jakubowska M, Kowalski Z (2006) Rapid pretreatment of a solid silver electrode for routine analytical practice. Electroanal 18(17):1710–1717
5. Fan CH, Li GX, Zhuang Y, Zhu JQ, Zhu DX (2000) Iodide modified silver electrode and its application to the electroanalysis of hemoglobin. Electroanal 12(3):205–208

6. Gutes A, Carraro C, Maboudian R (2011) Nonenzymatic glucose sensing based on deposited palladium nanoparticles on epoxy-silver electrodes. Electrochim Acta 56(17):5855–5859
7. Li GX, Liao XM, Fang HQ, Chen HY (1994) Direct electron-transfer reaction of hemoglobin at the bare silver electrode. J Electroanal Chem 369(1–2):267–269
8. Li GX, Chen HY, Zhu DX (1996) Imidazole modified silver electrode and its application to the investigation of the electrochemistry of cytochrome c. Anal Chim Acta 319(3):275–276
9. Eddowes MJ, Hill HAO (1977) Novel method for investigation of electrochemistry of metalloproteins—cytochrome-c. J Chem Soc, Chem Commun 21:771–772
10. Colon LA, Dadoo R, Zare RN (1993) Determination of carbohydrates by capillary zone electrophoresis with amperometric detection at a copper microelectrode. Anal Chem 65(4):476–481
11. Wasmus S, Kuver A (1999) Methanol oxidation and direct methanol fuel cells: a selective review. J Electroanal Chem 461(1–2):14–31
12. Armstrong FA, Hill HAO, Oliver BN (1984) Surface selectivity in the direct electrochemistry of redox proteins—contrasting behavior at edge and basal planes of graphite. J Chem Soc, Chem Commun 15:976–977
13. Hagen WR (1989) Direct electron-transfer of redox proteins at the bare glassy-carbon electrode. Eur J Biochem 182(3):523–530
14. Lane RF, Hubbard AT (1973) Electrochemistry of chemisorbed molecules. 1. Reactants connected to electrodes through olefinic substituents. J Phys Chem 77(11):1401–1410
15. Lane RF, Hubbard AT (1973) Electrochemistry of chemisorbed molecules. 2. Influence of charged chemisorbed molecules on electrode-reactions of platinum complexes. J Phys Chem 77(11):1411–1421
16. Herne TM, Tarlov MJ (1997) Characterization of DNA probes immobilized on gold surfaces. J Am Chem Soc 119(38):8916–8920
17. Chiang CK, Fincher CR, Park YW, Heeger AJ, Shirakawa H, Louis EJ, Gau SC, Macdiarmid AG (1977) Electrical-conductivity in doped polyacetylene. Phys Rev Lett 39(17):1098–1101
18. Liu HQ, Kameoka J, Czaplewski DA, Craighead HG (2004) Polymeric nanowire chemical sensor. Nano Lett 4(4):671–675
19. Saini S, Hall GF, Downs MEA, Turner APF (1991) Organic-phase enzyme electrodes. Anal Chim Acta 249(1):1–15
20. Bredas JL, Chance RR, Silbey R (1982) Comparative theoretical-study of the doping of conjugated polymers—polarons in polyacetylene and polyparaphenylene. Phys Rev B 26(10):5843–5854
21. Mandler D, Turyan I (1996) Applications of self-assembled monolayers in electroanalytical chemistry. Electroanal 8(3):207–213
22. Zhong CJ, Porter MD (1995) Designing interfaces at the molecular-level. Anal Chem 67(23):A709–A715
23. Otero R, Rosei F, Besenbacher F (2006) Scanning tunneling microscopy manipulation of complex organic molecules on solid surfaces. Annu Rev Phys Chem 57:497–525
24. Tseng TC, Urban C, Wang Y, Otero R, Tait SL, Alcami M, Ecija D, Trelka M, Gallego JM, Lin N, Konuma M, Starke U, Nefedov A, Langner A, Woll C, Herranz MA, Martin F, Martin N, Kern K, Miranda R (2010) Charge-transfer-induced structural rearrangements at both sides of organic/metal interfaces. Nat Chem 2(5):374–379
25. Chua LL, Zaumseil J, Chang JF, Ou ECW, Ho PKH, Sirringhaus H, Friend RH (2005) General observation of n-type field-effect behaviour in organic semiconductors. Nature 434(7030):194–199
26. Dimitrakopoulos CD, Malenfant PRL (2002) Organic thin film transistors for large area electronics. Adv Mater 14(2):99–117
27. Ishii H, Sugiyama K, Ito E, Seki K (1999) Energy level alignment and interfacial electronic structures at organic metal and organic/organic interfaces. Adv Mater 11(8):605–625
28. Qu XG, Chou J, Lu TH, Dong SJ, Zhou CL, Cotton TM (1995) Promoter effect of halogen anions on the direct electrochemical reaction of cytochrome-c at gold electrodes. J Electroanal Chem 381(1–2):81–85

29. Balzani V (2005) Nanoscience and nanotechnology: a personal view of a chemist. Small 1(3):278–283

30. Cao MH, He XY, Chen J, Hu CW (2007) Self-assembled nickel hydroxide three-dimensional nanostructures: a nanomaterial for alkaline rechargeable batteries. Cryst Growth Des 7(1):170–174

31. Huang XH, El-Sayed IH, Qian W, El-Sayed MA (2007) Cancer cells assemble and align gold nanorods conjugated to antibodies to produce highly enhanced, sharp, and polarized surface Raman spectra: a potential cancer diagnostic marker. Nano Lett 7(6):1591–1597

32. Miao P, Liu L, Nie YJ, Li GX (2009) An electrochemical sensing strategy for ultrasensitive detection of glutathione by using two gold electrodes and two complementary oligonucleotides. Biosens Bioelectron 24(11):3347–3351

33. Rao CNR, Sood AK, Subrahmanyam KS, Govindaraj A (2009) Graphene: the new two-dimensional nanomaterial. Angew Chem Int Edit 48(42):7752–7777

34. Valiev R (2002) Materials science—Nanomaterial advantage. Nature 419(6910):887–889

35. Wang J (2005) Nanomaterial-based amplified transduction of biomolecular interactions. Small 1(11):1036–1043

36. Wang J, Wang LH, Liu XF, Liang ZQ, Song SP, Li WX, Li GX, Fan CH (2007) A gold nanoparticle-based aptamer target binding readout for ATP assay. Adv Mater 19(22):3943–3946

37. Zhang YZ, Huang L (2012) Label-free electrochemical DNA biosensor based on a glassy carbon electrode modified with gold nanoparticles, polythionine, and graphene. Microchim Acta 176(3–4):463–470

38. Zhou ND, Wang J, Chen T, Yu ZG, Li GX (2006) Enlargement of gold nanoparticles on the surface of a self-assembled monolayer modified electrode: a mode in biosensor design. Anal Chem 78(14):5227–5230

39. Zhu XL, Zhao J, Wu Y, Shen ZM, Li GX (2011) Fabrication of a highly sensitive aptasensor for potassium with a nicking endonuclease-assisted signal amplification strategy. Anal Chem 83(11):4085–4089

40. Zhu ZQ, Su YY, Li J, Li D, Zhang J, Song SP, Zhao Y, Li GX, Fan CH (2009) Highly sensitive electrochemical sensor for mercury(II) Ions by using a mercury-specific oligonucleotide probe and gold nanoparticle-based amplification. Anal Chem 81(18):7660–7666

41. Jang SG, Kramer EJ, Hawker CJ (2011) Controlled supramolecular assembly of micelle-like gold nanoparticles in PS-b-P2VP diblock copolymers via hydrogen bonding. J Am Chem Soc 133(42):16986–16996

42. Padamwar MN, Patole MS, Pokharkar VB (2011) Chitosan-reduced gold nanoparticles: a novel carrier for the preparation of spray-dried liposomes for topical delivery. J Liposome Res 21(4):324–332

43. Xu YY, Wang J, Cao Y, Li GX (2011) Gold nanoparticles based colorimetric assay of protein poly(ADP-ribosyl)ation. Analyst 136(10):2044–2046

44. Zhu XL, Li YX, Yang JH, Liang ZQ, Li GX (2010) Gold nanoparticle-based colorimetric assay of single-nucleotide polymorphism of triplex DNA. Biosens Bioelectron 25(9):2135–2139

45. Cox JA, Jaworski RK, Kulesza PJ (1991) Electroanalysis with electrodes modified by inorganic films. Electroanal 3(9):869–877

46. Zak J, Kuwana T (1983) Chemically modified electrodes and electrocatalysis. J Electroanal Chem 150(1–2):645–664

47. Degrand C, Miller LL (1980) An electrode modified with polymer-bound dopamine which catalyzes NADH oxidation. J Am Chem Soc 102(18):5728–5732

48. Jaegfeldt H, Torstensson ABC, Gorton LGO, Johansson G (1981) Catalytic-oxidation of reduced nicotinamide adenine-dinucleotide by graphite-electrodes modified with adsorbed aromatics containing catechol functionalities. Anal Chem 53(13):1979–1982

49. Jaegfeldt H, Kuwana T, Johansson G (1983) Electrochemical stability of catechols with a pyrene side-chain strongly adsorbed on graphite-electrodes for catalytic-oxidation of dihydronicotinamide adenine-dinucleotide. J Am Chem Soc 105(7):1805–1814

50. Facci J, Murray RW (1982) Binding of pentachloroiridite to plasma polymerized vinylpyridine films and electrocatalytic oxidation of ascorbic-acid. Anal Chem 54(4):772–777

51. Kuo KN, Murray RW (1982) Electrocatalysis with ferrocyanide electrostatically trapped in an alkylaminesiloxane polymer film on a Pt electrode. J Electroanal Chem 131:37–60
52. Li FB, Dong SJ (1987) The electrocatalytic oxidation of ascorbic-acid on prussian blue film modified electrodes. Electrochim Acta 32(10):1511–1513
53. Tse DCS, Kuwana T (1978) Electrocatalysis of dihydronicotinamide adenosine-diphosphate with quinones and modified quinone electrodes. Anal Chem 50(9):1315–1318
54. Bookbinder DC, Lewis NS, Wrighton MS (1981) Heterogeneous one-electron reduction of metal-containing biological molecules using molecular-hydrogen as the reductant— Synthesis and use of a surface-confined viologen redox mediator that equilibrates with hydrogen. J Am Chem Soc 103(25):7656–7659
55. Chao S, Robbins JL, Wrighton MS (1983) A new ferrocenophane surface derivatizing reagent for the preparation of nearly reversible electrodes for horse heart ferri/ferrocytochrome c: 2,3,4,5-tetramethyl-1-((dichlorosilyl)methyl)[2]-ferrocenophane. J Am Chem Soc 105(2):181–188
56. Lewis NS, Wrighton MS (1981) Electrochemical reduction of horse heart ferricytochrome c at chemically derivatized electrodes. Science 211(4497):944–947
57. Cosgrove M, Moody GJ, Thomas JDR (1989) Metal-oxide catalyst membrane electrodes for the determination of hydrogen-peroxide. Analyst 114(12):1627–1632
58. Gorton L (1985) A carbon electrode sputtered with palladium and gold for the amperometric detection of hydrogen-peroxide. Anal Chim Acta 178(2):247–253
59. Itaya K, Shoji N, Uchida I (1984) Catalysis of the reduction of molecular-oxygen to water at prussian blue modified electrodes. J Am Chem Soc 106(12):3423–3429
60. Taniguchi I, Matsushita K, Okamoto M, Collin JP, Sauvage JP (1990) Catalytic-oxidation of hydrogen-peroxide at Ni cyclam modified electrodes and its application to the preparation of an amperometric glucose sensor. J Electroanal Chem 280(1):221–226
61. Hench LL, West JK (1990) The sol-gel process. Chem Rev 90(1):33–72
62. Avnir D (1995) Organic-chemistry within ceramic matrices-doped sol-gel materials. Accounts Chem Res 28(8):328–334
63. Gupta R, Chaudhury NK (2007) Entrapment of biomolecules in sol-gel matrix for applications in biosensors: problems and future prospects. Biosens Bioelectron 22(11):2387–2399
64. Cai WY, Xu Q, Zhao XN, Zhu JH, Chen HY (2006) Porous gold-nanoparticle-$CaCO_3$ hybrid material: preparation, characterization, and application for horseradish peroxidase assembly and direct electrochemistry. Chem Mater 18(2):279–284
65. Kang XH, Wang J, Tang ZW, Wu H, Lin YH (2009) Direct electrochemistry and electrocatalysis of horseradish peroxidase immobilized in hybrid organic-inorganic film of chitosan/sol-gel/carbon nanotubes. Talanta 78(1):120–125
66. Wang Y, Wu Y, Wang JW, Di JW (2009) Disposable superoxide anion biosensor based on superoxide dismutase entrapped in silica sol-gel matrix at gold nanoparticles modified ITO electrode. Bioproc Biosyst Eng 32(4):531–536
67. Xu JZ, Zhang Y, Li GX, Zhu JJ (2004) An electrochemical biosensor constructed by nano-sized silver particles doped sol-gel film. Mat Sci Eng C-Bio S 24(6–8):833–836
68. Bard AJ, Faulkner LR (2000) Electrochemical methods: fundamentals and applications, 2nd edn. Wiley, New York
69. Zoski CG (2007) Handbook of electrochemistry. Elsevier Science, The Netherlands

Chapter 3
Electrochemical Analysis of Proteins

Abstract Protein is of great importance to the execution of normal physiological functions of living organisms. Currently, the main techniques for the analysis of proteins include spectrophotometry, mass spectrometry, electrochemistry, affinity chromatography and enzyme-linked immunosorbent assay (ELISA), etc. Among them, electrochemical technique receives more and more interests. Especially in recent years, with the continuous improvement of protein-film voltammetry and the recently developed surface modification technology, nanotechnology, signal amplification technology and molecule recognition technology, electrochemical technique has been gradually overcoming the past drawbacks such as narrow research objects and sole signal format in the analysis of proteins. In this chapter, we review some typical strategies for the electrochemical analysis of protein activity as well as the electrochemical quantitative analysis of some proteins.

Keywords Protein activity • Quantitative analysis • Protein-film voltammetry • Electrochemical immunosensors • Electrochemical aptasensor • Enzyme catalysis • Nanomaterials-based amplification

3.1 Protein-Film Voltammetry

Protein-film voltammetry (PFV), proposed by Armstrong and coworkers, is a very useful methodology to obtain the direct electrochemistry of proteins, which can also characterize the mechanism how electron transfer process occurs at active sites [1, 2]. In this strategy, a film containing redox proteins is adsorbed on the surface of an electrode, where the proteins are usually immobilized on the electrode surface with appropriate orientation, electrons can be transferred in and out of the active sites of proteins by applying an appropriate potential.

Although many proteins exist on the interfaces in biological systems such as biological membrane, they are usually not able to take direct electron transfer reactions at the surface of bare electrode or even CMEs. Up to now, PFV is the best way to solve this problem. It also has many advantages over the other methodologies to obtain the direct electrochemistry of proteins [3]. As shown

Fig. 3.1 Schematic
illustration of the idealized
configuration for protein-film
voltammetry. The protein
molecules adsorbed on the
electrode form a perfect
monolayer. Reprinted
from Ref. [3], with kind
permission from Springer
Science + Business Media

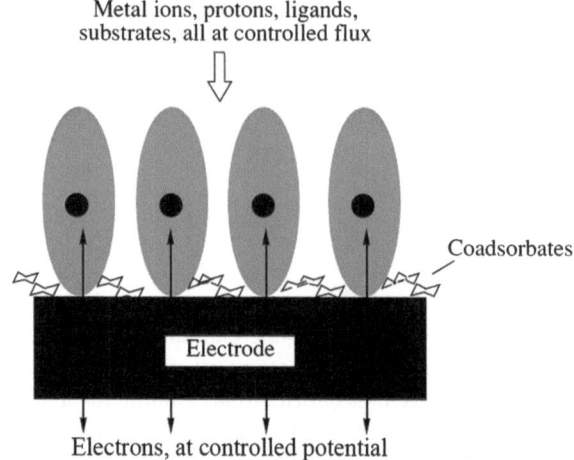

in Fig. 3.1, the proteins to be investigated can be configured on a suitable electrode as a relatively stable monolayer film, each of them behaves independently. Meanwhile, the proteins can form favorable orientations by interacting with functional groups on the electrode surface, which can facilitate the interfacial electron exchange and interaction. Moreover, the limitation due to diffusion of irregular macromolecules to an electrode surface can also be overcome. Besides, electron transfer can be realized by the modulation of the parameters such as potential and time; thus, high sensitivity can be achieved. Another advantage of PFV is that only very small protein samples are required for the studies. Furthermore, slow scan rate or potential step techniques can be used to study steady-state catalysis and redox-linked activation. Therefore, as the result of proliferating cross-disciplinary interactions, involving material science, surface science, analytical science, biochemistry, biophysics and so on, PFV has been developing very rapidly [4–6]. For the analysis of proteins, PFV has also been widely used. Table 3.1 summarizes the electrochemical analysis of some representative proteins by PFV.

Cyt *c* is one of the most studied analytes in protein analysis. Another reason why it is extensively studied with electrochemical technique is that Cyt *c* is among the early crystallized proteins. Meanwhile, it is a small heme protein containing just one heme group, and it works as an electron carrier in biological respiration. Cyt *c* may show excellent electron transfer activity not only in vivo but also in vitro. Therefore, as a redox protein, cyt *c* has been studied as a model for a myriad of electron transfer via different strategies [27, 28]. Hill and his coworker firstly established the reversible voltammetry of cyt *c* at a solid electrode [29]. After a controlled electrochemical potential was applied to the electrochemical system they proposed, which might be either continuous or in steps and pulses, well-defined redox waves could be observed. The obtained current response could also provide the thermodynamic and kinetic information of the protein. Later on, more and more electrochemical studies have been conducted on cyt *c*, including the quantitative analysis of the protein. For instance, Zhao et al. [11] have made

Table 3.1 Electrochemical analysis of some representative proteins by PFV

Protein	Electrode	Surface modification	References
Arsenite oxidase	PGE	None	[7]
Bilirubin oxidase	GCE	Nafion and CNTs	[8]
Blue copper proteins (parsley plastocyanin, azurin)	Boron-doped diamond	None	[9]
Cytochrome P450 reductase	PGE	Didodecyldimethylammonium bromide, dimyristoylphosphatidyl choline and poly(diallyldimethylammonium)	[10]
Cyt *c*	Au	Single-strand DNA-functionalized AuNPs	[11]
Cyt *c*	PGE	None	[12]
Cyt *c*	PGE	Agarose hydrogel	[13]
Diheme cyt c peroxidase	PGE	None	[14]
Hemoglobin	Ionic liquid modified carbon paste electrode	Nafion and ZnO nanoparticles	[15]
Hemoglobin	GCE	Microbial excocellular polysaccharide-gellan gum and ionic liquid 1-butyl-3-methyl-imidazolium hexafluorophosphate	[16]
Horseradish peroxidase(HRP)	GCE	Chitosan room temperature ionic liquid 1-butyl-3-methylimidazolium tetrafluoroborate film	[17]
HRP	PGE	Titanium dioxide nanoparticles	[18]
[NiFe]-Hydrogenase	Lithographically fabricated Au nano-electrode	Polymyxin-p	[19]
Microperoxidase	Tin-doped indium oxide on glass	Chitosan and CNTs	[20]
Microperoxidase	GCE	MWCNTs and AuNPs	[21]
Myoglobin	Au	MWCNTs and AuNPs	[22]
Myoglobin	GCE	Arylhydroxylamine	[23]
Myoglobin	PGE	SiO$_2$ nanoparticles and poly(diallyldimethylammonium)	[24]
Nitrite reductase	PGE	None	[25]
Human serum transferrin	PGE	Polyethyleneimine	[26]

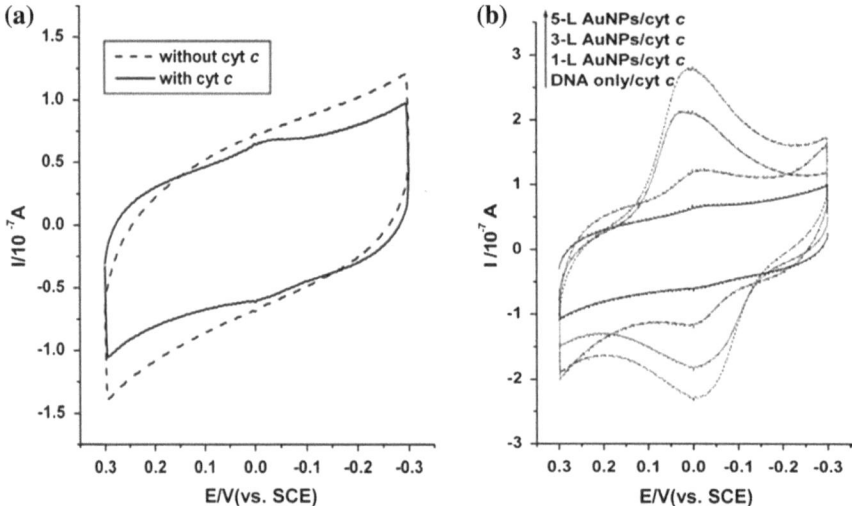

Fig. 3.2 Cyclic voltammograms obtained at HS-DNA-modified gold electrode **a** with and without the adsorption of cyt *c*; **b** with the adsorption of cyt *c* and the immobilization of different layers of AuNPs. Reproduced from Ref. [11] by permission of The Royal Society of Chemistry

use of the techniques of self-assembly and layer by layer, together with nanotechnology for the immobilization of cyt *c* on an electrode surface. In this strategy, single-strand DNA (ssDNA)-functionalized AuNPs were synthesized as the scaffolds for the construction of multilayered structure based on the recognition of complementary DNA strands. After the formation of uniformly self-assembled structure, cyt *c* can adsorb onto the electrode with native conformation. As shown in Fig. 3.2, the peak current of cyclic voltammograms increases with the number of AuNPs layers, demonstrating that the layers could not only provide a mild and compatible microenvironment for cyt *c* to execute the direct electron transfer reaction, but also greatly enhance the electrochemical signal due to the increase in the protein-binding sites. Therefore, stable and ultrasensitive electrochemical analysis of cyt *c* was achieved with a linear range from 2×10^{-9} to 1×10^{-7} M and a detection limit of 6.7×10^{-10} M.

Hemoglobin is another extensively studied protein. It makes up to 97 % of the red cell's dry content and about 35 % of the total content. Hemoglobin is significant for the storage and transport of oxygen in mammalian blood [30, 31]. When the ferrous ion in the protein is oxidized to ferric ion, the corresponding ferrihemoglobin (methemoglobin) cannot bind with oxygen; thus, the oxygen transport function is hampered. Nevertheless, methemoglobin can also be converted to active hemoglobin in some in vivo redox process such as the erythrocyte methemoglobin reduction pathway [32]. Compared with small heme proteins like cyt *c*, the direct electron transfer between hemoglobin and electrode is much more difficult to be obtained, because the heme groups of hemoglobin are deeply buried in their three-dimensional structure. Therefore, a lot of strategies have been proposed to improve the heterogeneous electron transfer [16, 33, 34]. Till now, many mediators have been employed to accelerate the electron transfer, and a great number of materials have been developed as film materials of PFV for electrochemical studies of hemoglobin, such as linoleic acid [35], DNA [36], poly-3-hydroxybutyrate [37], dipalmitoylphosphatidic acid [38], polyethylene glycol (PEG) [39], lactobionic acid [40], polyurethane elastomer [41], poly(ε-caprolactone) [42], mesoporous Al_2O_3 [43] and ionic liquid-TiO_2-graphene nanocomposite [44].

Human serum transferrin (hTf) is a typical iron transport protein which controls the level of free iron in biological fluids. It can also bind and transport Ti(IV) in blood. Two iron-binding sites of hTf can be occupied by Ti(IV), and the formed Ti(IV)-protein complex (Ti_2-hTf) can then be internalized into endosomes through transferrin receptor-mediated endocytosis. Shen et al. [26] have studied Ti_2-hTf through an electrochemical method. They found that polyethyleneimine could facilitate the electron transfer between PGE and Ti_2-hTf. The polymer could also maintain the native structure of Ti_2-hTf, behaving as a stabilizer. Therefore, direct electrochemistry of Ti_2-hTf was successfully achieved, and electrochemical analysis of the protein was conducted by using a polyethyleneimine film-modified electrode.

3.2 Electrochemical Immunosensors

Molecular recognition plays an essential role in biological systems. The molecular pairs include antibody and antigen, aptamer and target, enzyme and substrate, and receptor and ligand, all of which have been widely applied in protein analysis based on the fabrication of electrochemical immunosensors.

Antigen and its corresponding antibody, which may show high immunochemical affinity, is the most commonly used pair for the fabrication of electrochemical immunosensors. Antibody, also named as immunoglobulin, is a kind of relatively large protein produced by B cells to identify foreign objects including proteins, bacteria and viruses. These recognized objects are termed as antigens. Antibody consists of basic structural units of two heavy chains and two light chains. A small region at the tip of antibody is extremely variable, leading to millions of different antibodies with their unique tip structures for antigens to bind. On the other hand, antibody and antigen pair also has vital roles for the diagnosis of certain diseases, since many immunodiagnostic methods to diagnose infectious diseases, such as ELISA, immunofluorescence, Western blot and electrochemical immunosensors, are proposed based on the high affinity of this pair.

Electrochemical immunosensors, which may have not only high specificity but also high sensitivity, are very useful for protein analysis as well as the diagnoses of the related diseases. For example, the detection of anti-gliadin antibodies from human serum samples is essential for the diagnosis of autoimmune diseases like celiac disease, so an electrochemical immunosensor has been fabricated for the detection of anti-gliadin antibodies, which mimics traditional ELISA architecture [45]. Firstly, an alkane chain bipodal thiol SAM containing polyethylene glycol groups is modified

Fig. 3.3 The calibration curves of immunoassay of CEA: **a** GCE/Au-TiO$_2$/ Ab$_1$/BSA/CEA/HRP-Ab$_2$- HPtNPs; **b** GCE/Au-TiO$_2$/ Ab$_1$/BSA/CEA/HRP-Ab$_2$- SPtNPs; **c** GCE/Au-TiO$_2$/ Ab$_1$/BSA/CEA/HRP-Ab$_2$. The inset shows cyclic voltammograms of GCE/Au-TiO$_2$/ Ab$_1$/BSA/CEA/HRP-Ab$_2$- HPtNPs with various CEA concentrations. Reprinted from Ref. [47], Copyright 2011, with permission from Elsevier

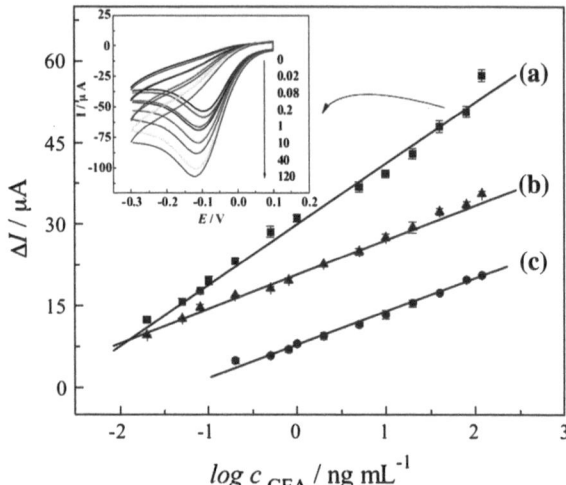

on a gold electrode for further conjugation of the antigen gliadin via carbodiimide chemistry. Then, the bipodal nature of the alkane thiol preserves a good diffusion of electroactive species toward the electrode, which can also improve the stability of the monolayer. Therefore, amperometric evaluation of the immunosensor for the detection of anti-gliadin antibodies can show stable and reproducible low limit of detection (46 ng/mL; RSD = 8.2 %, n = 5), which is attributed to the highly sensitive inherence of electrochemical measurement. The biosensor also allows the estimation of semi-quantitative antibody contents in thirty-minute total assay time without the interference of complex matrices.

The p53 protein is a kind of tumor suppressor protein with many functions and binding partners in cells. It acts as a cell guardian in cellular proliferation. Therefore, Yeo et al. [46] have developed a selective platform for electrochemically monitoring the cellular p53 concentration. They firstly immobilize the antibody molecules on an (R)-lipo-diaza-18-crown-6 SAM-modified gold electrode. The capture of p53 protein can then modulate the charge transfer, which can be monitored by electrochemical impedance spectroscopy. This selective detection of p53 protein can also be used to indicate cancers and other cellular malfunctions.

Carcinoembryonic antigen (CEA) is a glycoprotein involved in cell adhesion, which is a kind of broad-spectrum tumor marker. Recently, a sensitive electrochemical immunosensor for the determination of CEA was fabricated [47]. In this work, Au–TiO$_2$ nanoparticles and multiple HRP-labeled antibodies-functionalized hollow platinum nanospheres (HPtNPs) (abbreviated as HRP–Ab$_2$–HPtNPs) are employed. Au–TiO$_2$ nanoparticles are firstly modified on GCE before the loading of anti–CEA antibody and the further blocking by bovine serum albumin (BSA). Therefore, CEA to be detected can then bring HRP–Ab$_2$–HPtNPs onto the electrode surface. After the sandwich immunoreaction occurs, the reduction of H$_2$O$_2$ catalyzed by HRP will generate electrochemical response, in which hydroquinone (H$_2$Q) is used as the mediator. Figure 3.3 shows the detection of a standard CEA solution by cyclic voltammetry. With the increased concentration of CEA, the current is increased gradually. The results suggest that the detection is highly sensitive, which is attributed to the signal amplification from the synergistic action of HPtNPs and HRP.

Antibodies have also been widely used in protein analysis with clinical importance, for instance, as important biomarkers for the diagnosis of infection and many autoimmune diseases, since the need for point-of-care antibody measurement increases greatly in recent years, which may benefit the diagnosis, prevention and treatment of many illnesses [48–51]. As a good example, we would introduce a wash-free, electrochemical platform for multiplexed detection of specific antibodies proposed by White et al. [52]. In this study, a nucleic acid duplex for molecular recognition and signal generation is designed (Fig. 3.4). While one strand is affixed to electrode via its 5′ terminus, whose 3′ terminus is modified with a methylene blue to produce electrochemical signals, the other strand is modified with antigen at its 5′ terminus to recognize target antibody. Since the immune reaction can prevent methylene blue approach the electrode surface, efficient electron transfer is blocked. Therefore, the amount of binding antibody can lead to the readily measureable change in faradaic current, recorded by square wave voltammetry, and a nanomolar detection limit is sufficiently enough for many applications.

Electrochemical immunosensors have been applied as well in the protein analysis for environmental monitoring. For instance, Chen et al. [53] have reported a highly sensitive and selective electrochemical assay of acetylcholinesterase (AChE) activity in red blood cells, which is also a useful biomarker to monitor the exposures to organophosphorus (OP) pesticides and chemical nerve agents. In this strategy, MWCNTs–Au nanocomposites-modified screen-printed carbon electrode is firstly constructed for the immobilization of AChE-specific antibody. The target AChE is then captured on the electrode surface by the immunoreaction. Electrochemical detection of AChE can be executed over a linear range from 0.1 to 10 nM. This assay method has been further applied for monitoring the OP exposure, since the OP-AChE composite may inhibit the immunoreaction (Fig. 3.5). Experimental results show that less than 5 % inhibition can be detected, which promises the immunosensor a quite selective and sensitive assay of AChE activity for biomonitoring of exposure to OP pesticides.

3.3 Aptamer-Based Electrochemical Analysis of Proteins

Aptamers are single-strand oligonucleic acid ligands that bind to a specific target molecule or cell with high binding affinity and specificity [54, 55]. The inherent flexibility makes aptamers fold into specific tertiary structures after binding to their targets. In 1991, the technology of systematic evolution of ligand by exponential enrichment (SELEX) was developed for isolation of aptamers [56]. Since then, more and more aptamers have been screened for binding to various

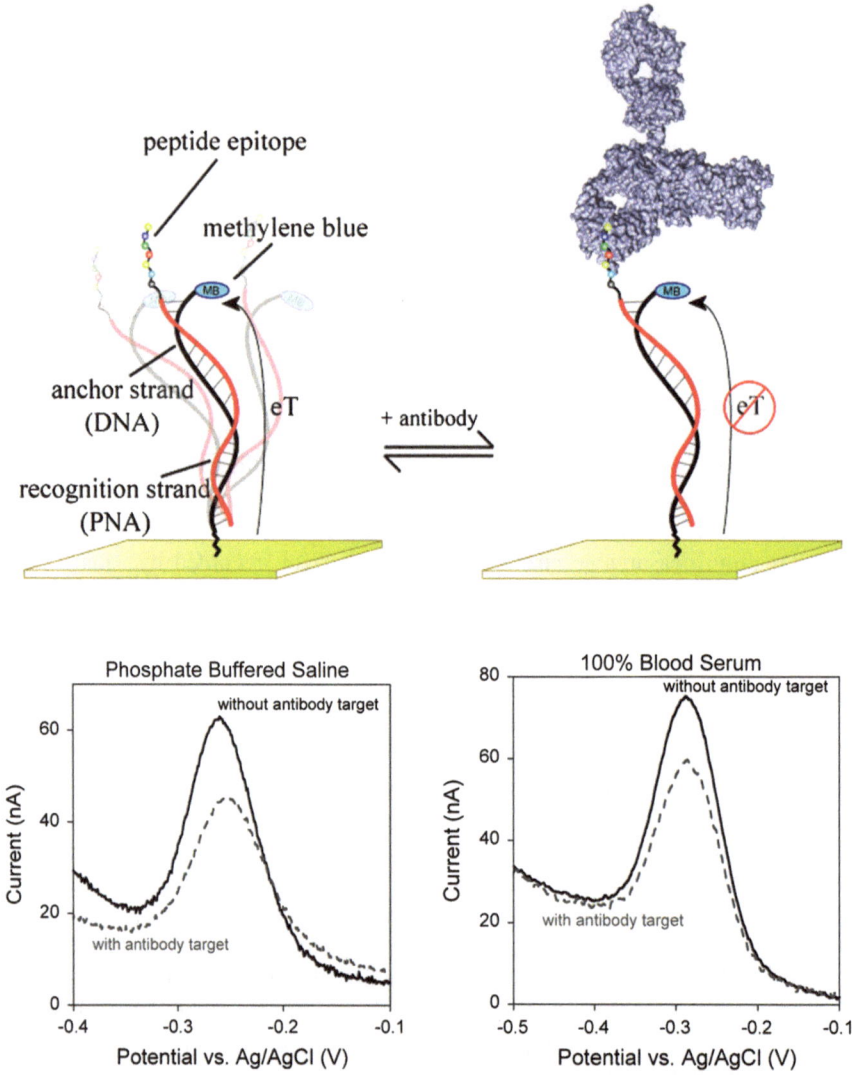

Fig. 3.4 Schematic illustration of the electrochemical detection of specific antibodies (*top*). Without the immuno binding of antibody (*top left*), the surface attachment chemistry is flexible, facilitating efficient electron transfer between the redox reporter and the electrode surface. With the target antibody binding (*top right*), electron transfer decreases, presumably due to the decrease of the efficiency of the reporter collides with the electrode. The immuno binding can be monitored from the change in peak current of square wave voltammetry (*Bottom*). Reprinted with the permission from Ref. [52]. Copyright 2012 American Chemical Society

targets including ions, small molecules, proteins and even cells. They are recognized as "chemical antibodies" which have lots of merits in the application for electrochemical analysis of proteins and cells [57–59]. Firstly, chemical synthesis

Fig. 3.5 Schematic illustration of the electrochemical analysis of acetylcholinesterase activity and its exposures to organophosphorus pesticides. Reprinted with the permission from Ref. [53]. Copyright 2012 American Chemical Society

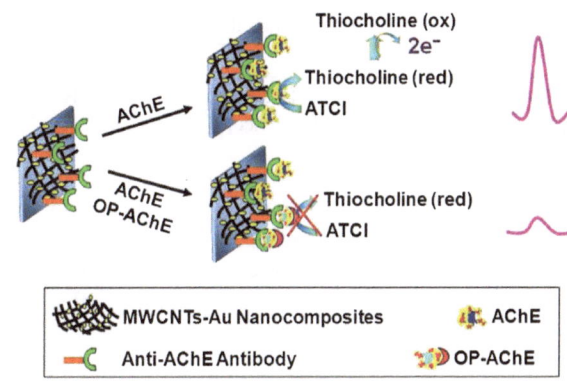

of aptamers is quite easy and cost-effective with superb stability in harsh conditions [60]. Secondly, aptamers against targets can be selected in either physiological or non-physiological conditions, which greatly facilitates the selection [61–63]. Thirdly, aptamers show high affinity and specificity for a variety of targets; thus, the applications of corresponding analysis are broad [64–66]. Finally, aptamers can be chemically modified more easily than antibodies, especially for the modification of signal moieties, like electrochemical probes. These features may facilitate the realization of many strategies for electrochemical analysis. Moreover, aptamers can hybridize with their complementary sequences. This nature can be used to design versatile oligonucleotides machines for either protein/cell analysis or clinical applications [67–69]. In addition, the incorporating electrochemical molecular reporters can make the analysis systems more sensitive, where methylene blue, ferrocene and pentaamminechlororuthenium (III) chloride ($[Ru(NH_3)_5Cl]^{2+}$) are the frequently used electrochemical reporters.

Thrombin is a "trypsin-like" serine protease, which converts soluble fibrinogen into insoluble strands of fibrin. It may induce thrombosis and tumor angiogenesis. It can also catalyze many other coagulation-related reactions. So, the study of this enzyme has attracted much attention [70, 71]. So far, a great many strategies for thrombin analysis have been developed. Table 3.2 summarizes the detection limit and the employed techniques for the analysis of thrombin.

Xiao et al. [76] once proposed an electrochemical signal-off strategy to detect thrombin. As shown in Fig. 3.6, before thrombin is bound to its aptamer, methylene blue is firstly covalently labeled at the end of the aptamer, so electrochemical signal can be obtained at the aptamer-modified electrode. Nevertheless, upon the recognition of thrombin with its aptamer immobilized on the electrode surface, G-quadruplex conformation will be formed; thus, the methylene blue molecule attached at the end of the aptamer is shielded from the electrode, and consequently, no electrochemical signal can be obtained.

The above strategy has a main drawback of negative signal. In order to overcome this disadvantage, some colleagues have proposed some signal-on methods. For instance, Radi et al. [74] have prepared a polycrystalline gold electrode modified with anti-thrombin aptamer. At the other end of the aptamer, ferrocene

Table 3.2 Overview of aptamer-based thrombin analysis

Detection methods	Detection limit (M)	References
Electrochemical impedance spectroscopy	2.0×10^{-9}	[72]
Electrochemical impedance spectroscopy	1.0×10^{-13}	[73]
DPV	5×10^{-10}	[74]
DPV	5.5×10^{-16}	[75]
Alternating current voltammetry	6.4×10^{-9}	[76]
Square wave stripping voltammograms	5×10^{-13}	[77]
Enhanced electrochemiluminescence	5×10^{-14}	[78]
Raman spectroscopy	1×10^{-10}	[79]
Surface acoustic wave	4×10^{-10}	[80]
Fluorescence	1.06×10^{-9}	[81]
qPCR	4.5×10^{-13}	[82]

Fig. 3.6 Schematic representation of the design of the analysis of thrombin by a signal-off electrochemical strategy. *MB* = methylene blue. Reproduced from Ref. [76] by permission of John Wiley and Sons Ltd

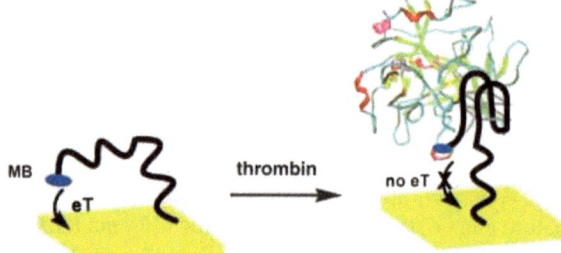

is covalently attached as the electrochemical reporter, and 2-mercaptoethanol is used to further block the electrode surface. Before the binding of the aptamer with thrombin, no signal can be obtained due to the large electron-tunneling distance between ferrocene moiety and the electrode. However, after the binding with thrombin, G-quadruplex structure is formed and the ferrocene moiety is dragged to approach the electrode surface, generating significant electrochemical signals.

Since thrombin has dual binding sites to its aptamer, a "sandwich" structure can be fabricated for the analysis of this protein. For example, Deng et al. [73] have developed an impedimetric aptasensor with femtomolar sensitivity based on the enlargement of surface-charged AuNPs. While one aptamer immobilized on a gold electrode surface can capture thrombin, another aptamer which is modified on the surface of AuNPs can also bind with thrombin to form a "sandwich" structure. Since the changes of resistance can be probed by $Fe(CN)_6^{3-/4-}$, the electron transfer resistance of the electrode may indicate the amount of captured thrombin. Compared with a previously reported work [72], a femtomolar sensitivity can be achieved, because the impedimetric response obtain by this aptasensor has been greatly enhanced with the help of the surface-charged AuNPs.

Many other label-free impedimetric aptasensors have also been developed based on the conductivity changes induced from the aptamer–target interactions.

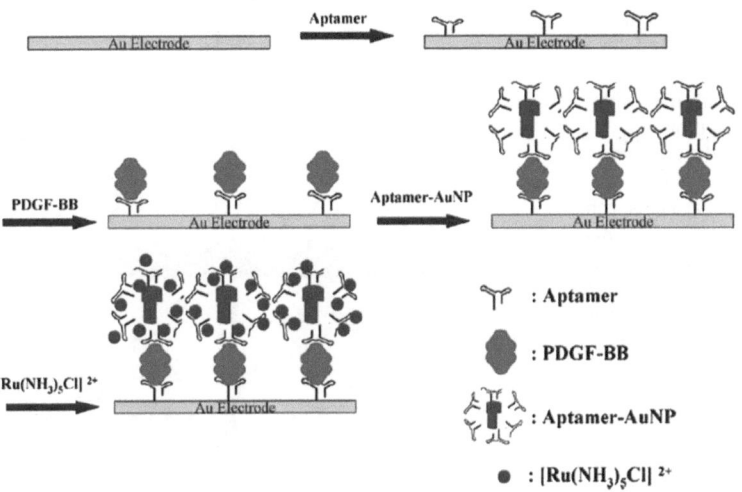

Fig. 3.7 Procedures of the preparation of the "Sandwich" Sensing System. Reprinted from Ref. [84], Copyright 2009, with permission from Elsevier

For example, Xu et al. [83] have reported a method to detect human IgE on the basis of the impedance changes attributed to the aptamer–target binding events. The detection range of human IgE by this aptamer-based impedance method lies in 2.5–100 nM with a detection limit as low as 0.1 nM. On the other hand, "sandwich" structures have also been fabricated for the analysis of some other proteins. Platelet-derived growth factor (PDGF) regulates cell growth and division and plays a significant role in blood vessel formation. The uncontrolled expression procedure can also be a symbol of cancer. Therefore, Wang et al. [84] have proposed an electrochemical method for PDGF analysis based on anti-PDGF aptamer and AuNPs. As shown in Fig. 3.7, the aptamer is firstly self-assembled on an electrode surface. In the presence of the target PDGF, PDGF-aptamer complex is formed with one binding site of PDGF being used. Then, another aptamer which is modified on the surface of AuNPs recognizes the other binding site of PDGF; thus, the sandwich structure is formed. In this assay, positive-charged $[Ru(NH_3)_5Cl]^{2+}$ can be adsorbed on the aptamers that are modified on the surface of AuNPs via electrostatic interaction; thus, amplification of electrochemical signal can be achieved. Meanwhile, AuNPs can also amplify the electrochemical signal. Consequently, a detection limit of 1×10^{-14} M for purified samples and 1×10^{-12} M for complex samples can be achieved.

Multiple protein analysis is always desired for the assay of proteins as well as clinical applications since the detection of a large number of disease markers with ultralow levels is highly required during early diagnosis of some diseases. Hansen et al. [77] have proposed an aptamer-based biosensor for multiple protein analysis coupled with enormous amplification feature of nanoparticle-based electrochemical stripping measurements. As shown in Fig. 3.8, several thiolated aptamers are modified on the electrode surface with the corresponding quantum dot-tagged proteins

Fig. 3.8 **a** Multiplied immobilization of thiolated aptamers on the gold substrate with the bound protein–quantum dot conjugates; **b** the displacement of the tagged proteins; **c** the electrochemical stripping detection at a coated glassy carbon electrode after the dissolution of the remaining captured nanocrystals. Reprinted with the permission from Ref. [77]. Copyright 2006 American Chemical Society

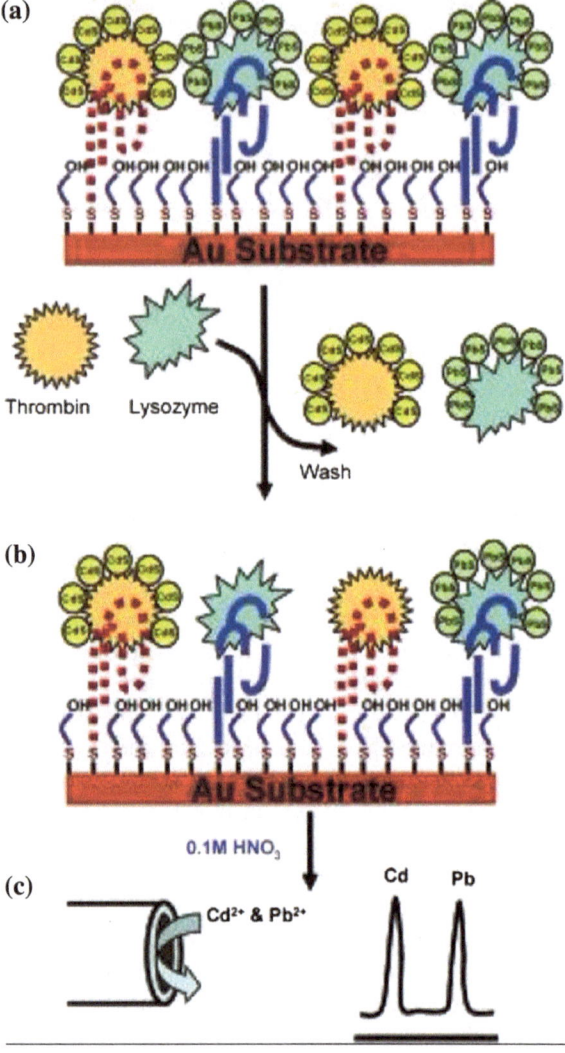

bound to them. The addition of target proteins would displace the corresponding quantum dot-tagged ones; thus, the remaining nanocrystals is electrochemically detected, which can be converted to the concentration of the target proteins.

3.4 Electrochemical Analysis Based on Enzyme Catalysis

A lot of proteins have enzymatic activity, which can catalyze corresponding substrate to generate detectable electrochemical signals. This process may not only be used for protein analysis, but also for the great enhancement of the sensitivity for protein detection.

Fig. 3.9 Schematic illustration of the strategy for the electrochemical detection of PTK activity coupled with electrocatalyzed Tyr oxidation as the signal reporter. Reprinted from Ref. [87], Copyright 2011, with permission from Elsevier

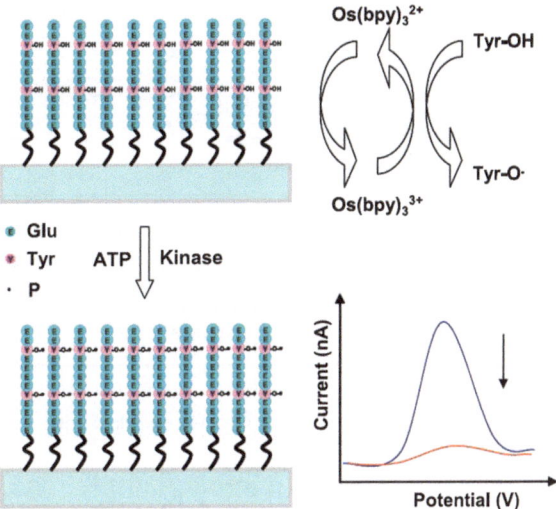

Protein kinase–catalyzed phosphorylation plays a significant role in many vital biological processes including cell signal transduction, cell proliferation and differentiation [85, 86]. Tyrosine kinases are a subclass of protein kinase, which can catalyze the reaction of transferring a phosphate group from ATP to a protein. Yang et al. [87] have made use of the catalytic activity of tyrosine kinases to develop a label-free electrochemical method to assay the activity of the enzyme. While poly (Glu, Tyr) (4:1) peptide is immobilized on the surface of an indium tin oxide (ITO) electrode as the substrate of epidermal growth factor receptor (EGFR), a typical tyrosine kinase, the tyrosine residue of the peptide can be used as an electrochemical signal reporter, since this residue can be also electrocatalyzed by a dissolved electron mediator $Os(bpy)_3^{2+}$ (bpy $= 2,2'$-bipyridine). With the phosphorylation of Tyr by tyrosine kinase, the electrochemical response is sharply decreased due to the loss of free phenol group of Tyr (Fig. 3.9), so the activity of this enzyme can be detected and the kinase inhibition can also be monitored by this electrochemical method.

Protein kinase A is a family of enzymes whose activity depends on the cellular levels of cyclic AMP (cAMP). This enzyme family regulates metabolism of glycogen, sugar and lipid. Miao et al. [88]. have developed an electrochemical strategy for sensing protein kinase A, based on Zr^{4+}-mediated signal transition and rolling circle signal amplification. Firstly, the substrate peptide is immobilized on a gold electrode surface through thiol group of cysteine residue. Then, another residue, serine of the peptide, is catalyzed by protein kinase A. After that, Zr^{4+} links the phosphorylated serine and a designed DNA primer probe via the interaction with the phosphate groups. Consequently, rolling circle amplification is initiated from the DNA primer probe, generating a long DNA chain, which can further adsorb a large number of electrochemical species, in token of the phosphorylation process by protein kinase A (Fig. 3.10).

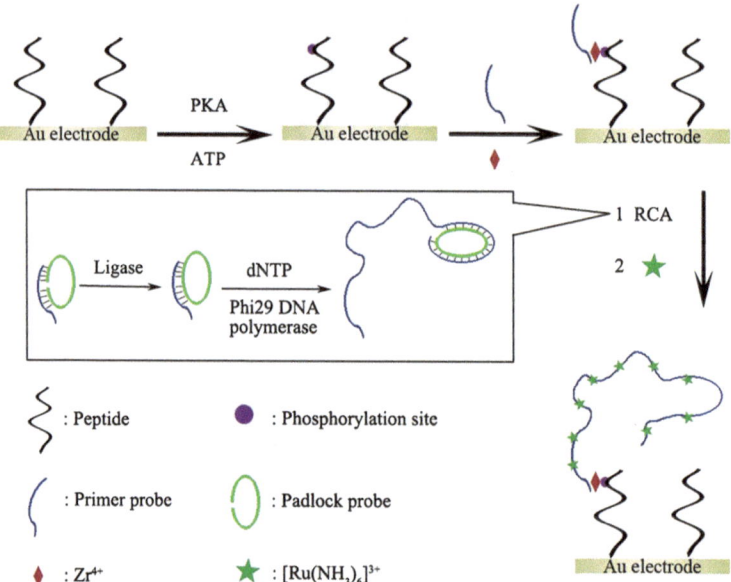

Fig. 3.10 Schematic representation of the electrochemical sensing strategy for the activity of protein kinase. Reprinted with the permission from Ref. [88]. Copyright 2012 American Chemical Society

This laboratory has also reported a switchable "On–Off" electrochemical strategy for the detection of protein kinase A, which can be re-operated for many times [89]. Firstly, a specific substrate peptide is immobilized on the surface of an electrode, forming a compact and positively charged SAM. Since the peptide is designed as electro-positive, the formed monolayer will block the positively charged electrochemical probes $[Ru(NH_3)_5Cl]^{2+}$ to get access to the electrode surface; thus, the detection system is set in the "Off" state. However, after the phosphorylation catalyzed by protein kinase A takes place, the SAM becomes loose. Due to the existence of phosphate groups, $[Ru(NH_3)_5Cl]^{2+}$ probes then get close to the electrode surface by electrostatic interaction, making the detection system lies in the "On" state. Figure 3.11 is the differential pulse voltammograms of $[Ru(NH_3)_5Cl]^{2+}$ to show the two states of the detection system. It also reveals that this detection system can be reused for many times without significant loss of detection precision.

In contrast with protein kinase, phosphatase is a kind of enzyme that removes a phosphate group from its substrate. This reaction is termed as dephosphorylation, which can be found in muscle movement and many other reactions within the body. Alkaline phosphatase is a common phosphatase in many organisms. Miao et al. [90] have proposed an electrochemical method for the detection of alkaline phosphatase by the use of two complementary DNA probes (DNA 1 and DNA 2) coupled with λ exonuclease. Briefly, the 5′ phosphoryl end of DNA 1 is

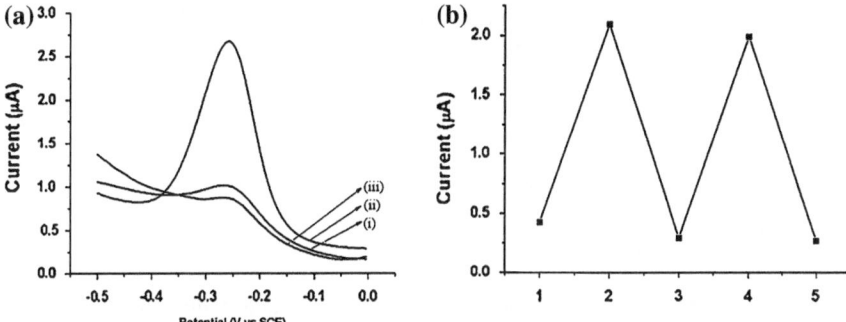

Fig. 3.11 **a** Differential pulse voltammograms of $[Ru(NH_3)_5Cl]^{2+}$ obtained at the substrate peptide-modified electrode before (*i*) and after (*ii*) the catalysis of 1-unit/mL PKA (*iii*) is the case that 200-units/mL ALP is further added into the test solution; **b** Relationship between peak current of the $[Ru(NH_3)_5Cl]^{2+}$ solution obtained at the substrate peptide-modified electrode and the repeated times of the phosphorylation and dephosphorylation. Reprinted from Ref. [89], Copyright 2010, with permission from Elsevier

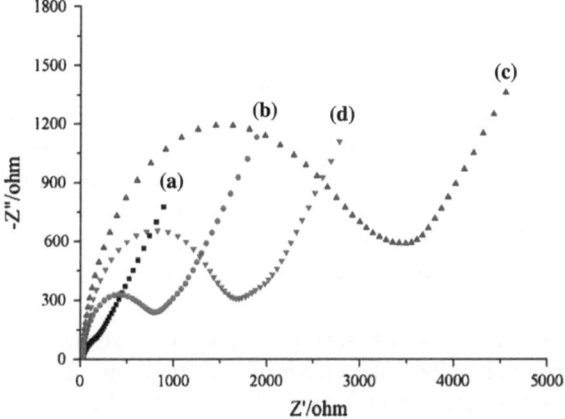

Fig. 3.12 Nyquist plot of impedance of **a** the bare gold electrode, **b** MCH-modified electrode, **c** DNA 2 and MCH-modified electrode, **d** DNA 2 and MCH-modified electrode after dephosphorylated DNA 1 and λ exonuclease were separately added in the test solution. Reprinted from Ref. [90], Copyright 2011, with permission from Elsevier

dephosphorylated by alkaline phosphatase. After DNA1 is hybridized with DNA 2, which has been previously immobilized on an electrode surface, DNA 2 will be cleaved by λ exonuclease owing to the phosphoryl of DNA 2 at the 5′ end. The digestion of DNA 2 can further release DNA 1 from the double strands; thus, DNA 1 will hybridize with another DNA 2 strand, initiating another digestion cycle. With the loss of large amount of DNA 2 and the adsorbed electrochemical species on the electrode surface, the electrochemical response is brought down. Figure 3.12 depicts the impedance change of the electrode surface in token of the process of dephosphorylation and the digestion cycles by λ exonuclease, which can also be shown by the measurements with cyclic voltammetry or chronocoulometry.

3.5 Nanomaterials-Based Electrochemical Analysis of Proteins

With the tremendous development in material science and nanotechnology, many nanomaterials with superior properties have been synthesized and further functionalized for the electrochemical analysis of proteins, which has also dramatically improved the detection performance [91–94]. Therefore, some metal nanoparticles, magnetic nanoparticles, nanowires and nanotubes have been employed either as electrical connectors between electrodes and the redox centers of proteins or as signal amplification elements for protein analysis [95, 96].

TiO_2 is a kind of n-type semiconductor material with excellent biocompatibility and environmental safety. In nanometer scale, TiO_2 may show a better biocompatibility, and the material will have a higher ratio of surface area to volume. These properties make this material more appropriate for protein analysis, so TiO_2 nanoparticles have received much attention from the colleagues in this community [97, 98]. For instance, with the help of TiO_2 nanoparticles, Zhou et al. [99] have investigated the catalytic ability of tyrosinase by electrochemical technique. Experimental results reveal that the photovoltaic effect of TiO_2 nanoparticles can make tyrosinase exhibit a higher enzyme activity. The reason might be that large amount of oxytyrosinase, which will act as an intermediate in the process of catalytic reactions, is generated due to the photovoltaic effect of TiO_2 under ultraviolet (UV) light irradiation.

Pt nanoparticles have also been proposed as ideal substitutes of enzymes to catalyze some reactions [100], the enzymatic reaction rate of which is even one to two orders of magnitude higher than that of HRP [101], a commonly used enzyme for biosensor fabrication. Cao et al. [102] have made use of this nature and reported a strategy to detect indoleamine 2,3-dioxygenase (IDO), a key protein in the inhibition of T-cell proliferation, synthesis of neurontransmitter and regulation of immune responses. The system of the electrochemical analysis combines the enzymatic catalysis and electrocatalysis of Pt nanoparticles, which can greatly improve the sensitivity of the detection. Under optimized conditions, this assay method can detect IDO activity in the range of 20–400 unit/mL with a detection limit of 6.84 unit/mL. The principle of this strategy can be illustrated in Fig. 3.13. Firstly, tryptophan, the substrate of IDO, is immobilized on a gold electrode surface. With the oxidation catalyzed by IDO, tryptophan is hydrolyzed and kynurenine is generated, which may induce the further immobilization of dithiobis [succinimidylpropionate]-modified Pt nanoparticles onto the electrode surface. Since Pt nanoparticles can electrochemically catalyze the reduction of H_2O_2, the produced electrochemical signal is then correlated with the enzyme activity of IDO.

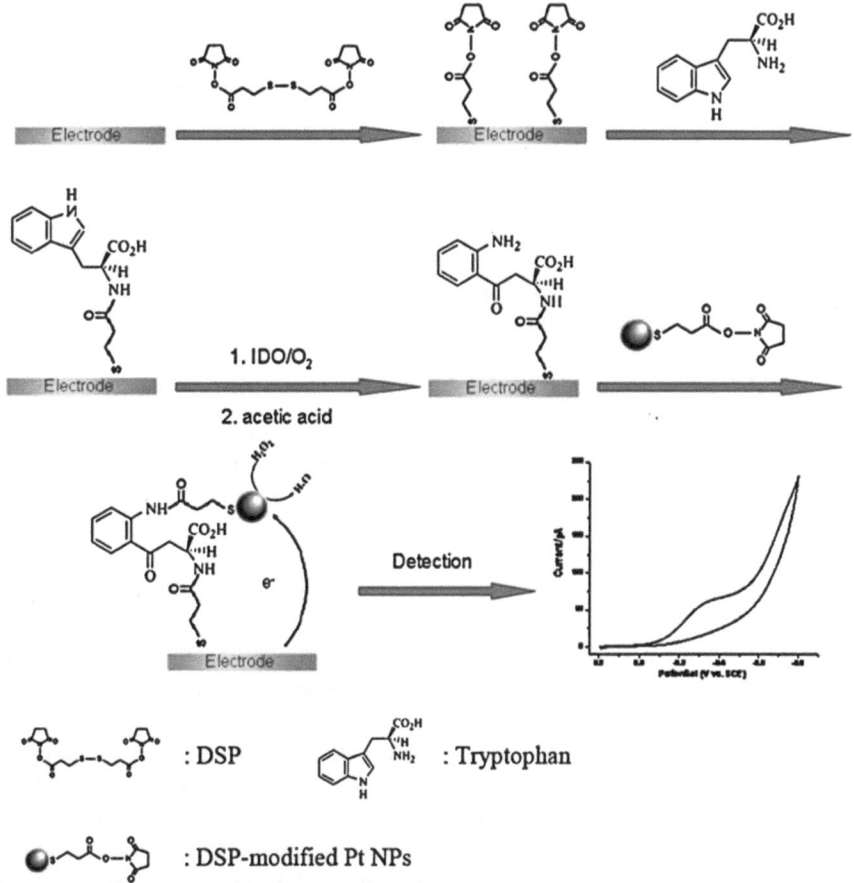

Fig. 3.13 Schematic representation of the electrochemical analysis of IDO activity. Reprinted from Ref. [102], Copyright 2010, with permission from Elsevier

Graphene is a newly developed advanced two-dimensional nanomaterial, discovered in the year of 2004 [103]. Graphene has many unique properties including fast electron transportation, high thermal conductivity, excellent biocompatibility, which promise its great application potential in electrochemical analysis of proteins. Zhao et al. [104] have reported a simple and smart electrochemical biosensing platform based on PGE modified with graphene quantum dots with the help of some specific ssDNA added in the test solution. Although the good conductivity of graphene material may result in large electrochemical response of the modified electrode, the single-strand DNA will be strongly bound to graphene material, inhibiting the electron transportation between $[Fe(CN)_6]^{3-/4-}$ and the substrate electrode. Nevertheless, when target protein also exists in the test solution, the ssDNA added in the test solution, which is designed as the aptamer of the target protein, will bind with its target, instead of binding

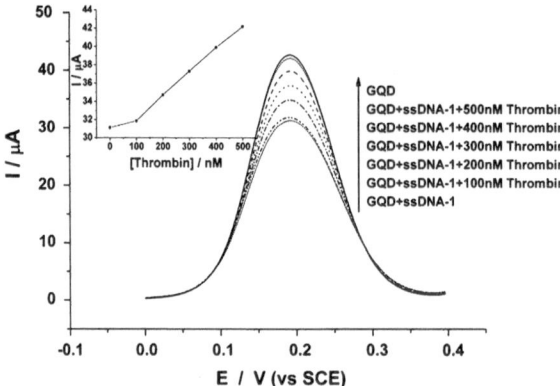

Fig. 3.14 Differential pulse voltammograms of $[Fe(CN)_6]^{3-/4-}$ obtained at a graphene quantum dots–modified electrode for the cases that ssDNA-1 has been incubated by thrombin with different concentrations. Insets show the plots of the peak current against the concentration of ssDNA-2 or thrombin. Reprinted from Ref. [104], Copyright 2011, with permission from Elsevier

with the graphene quantum dots modified on the PGE surface; thus, the peak current of $[Fe(CN)_6]^{3-/4-}$ is increased with the increased concentration of the target protein (Fig. 3.14).

AuNPs are the most popular nanomaterials with extensive range of useful properties like high surface area, localized surface plasmon resonance, superquenching capability and high extinction coefficients, etc. DNA-functionalized AuNPs are especially useful for the development of research strategies for protein analysis. Ding et al. [75] have prepared aptamer-functionalized AuNPs to carry smaller CdS nanoparticles which are used as signal reporter for the determination of thrombin. Figure 3.15 is the schematic illustration of their design. AuNPs are co-modified with anti-thrombin aptamer and a large number of CdS nanoparticles. Since thrombin to be detected can bind the aptamers modified on the surfaces of both the electrode and AuNPs, a "sandwich" structure is formed. Since a large amount of CdS nanoparticles are loaded on the surface of AuNPs, which may release a great number of Cd^{2+} ions, very high DPV signals are generated; thus, ultrasensitive detection of thrombin is achieved.

CNT is another commonly used material for protein analysis, which is a one-dimensional nanomaterial with long and slender structures. The unique hexagonal networks and C–C covalent bonding make it attractive in the fabrication of many electrochemical analysis systems. For instance, Wan et al. [105] have developed a multiplexing electrochemical immunosensor for the detection of cancer-related protein biomarkers. In this device, both HRP and goat anti-rabbit IgG (secondary antibody, Ab₂)–modified MWCNTs are prepared. As shown in Fig. 3.16, after the formation of "sandwich" structure, multichannel detection of a series of target biomarkers can be achieved.

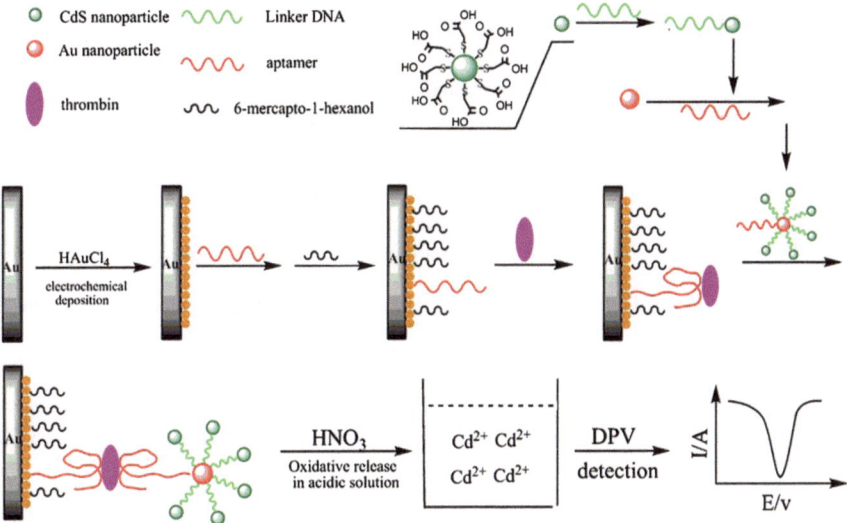

Fig. 3.15 Schematic demonstration for the electrochemical sensor of thrombin based on AuNPs and CdS nanoparticles. Reprinted from Ref. [75], Copyright 2010, with permission from Elsevier

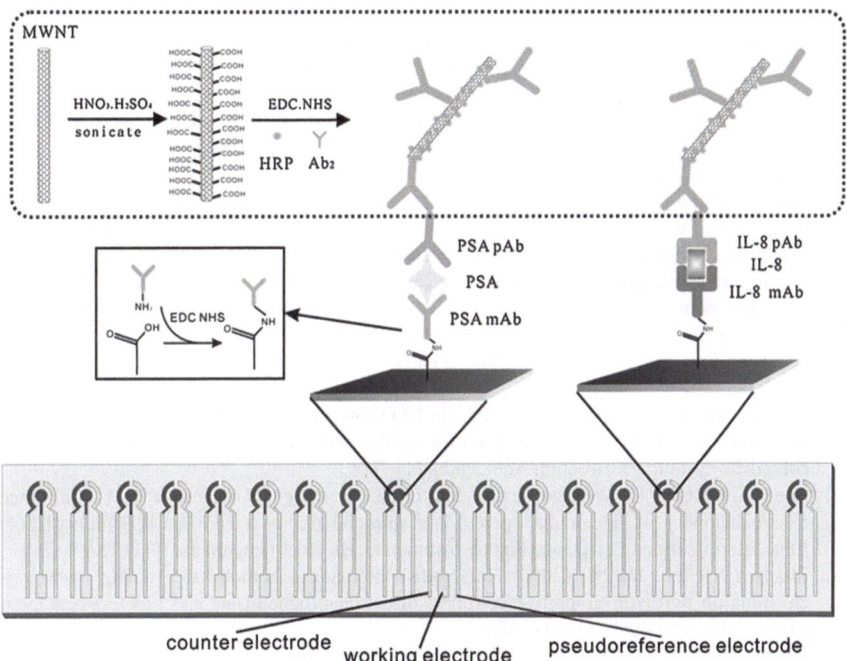

Fig. 3.16 Schematic representation of the "sandwich"-type strategy for electrochemical immunosensors detecting cancer biomarkers. Reprinted from Ref. [105], Copyright 2011, with permission from Elsevier

References

1. Armstrong FA (2002) Insights from protein film voltammetry into mechanisms of complex biological electron-transfer reactions. J Chem Soc Dalton 5:661–671
2. Armstrong FA, Heering HA, Hirst J (1997) Reactions of complex metalloproteins studied by protein-film voltammetry. Chem Soc Rev 26(3):169–179
3. Armstrong FA (2002) Protein film voltammetry: revealing the mechanisms of biological oxidation and reduction. Russ J Electrochem 38 (1):49–62
4. Zhang WJ, Li GX (2004) Third-generation biosensors based on the direct electron transfer of proteins. Anal Sci 20(4):603–609
5. Baffert C, Bertini L, Lautier T, Greco C, Sybirna K, Ezanno P, Etienne E, Soucaille P, Bertrand P, Bottin H, Meynial-Salles I, De Gioia L, Leger C (2011) CO disrupts the reduced h-cluster of FeFe hydrogenase. A combined DFT and protein film voltammetry study. J Am Chem Soc 133(7):2096–2099
6. Chen KS, Hirst J, Camba R, Bonagura CA, Stout CD, Burgess BK, Armstrong FA (2000) Atomically defined mechanism for proton transfer to a buried redox centre in a protein. Nature 405(6788):814–817
7. Bernhardt PV, Santini JM (2006) Protein film voltammetry of arsenite oxidase from the chemolithoautotrophic arsenite-oxidizing bacterium NT-26. Biochemistry 45(9):2804–2809
8. Ivnitski D, Artyushkova K, Atanassov P (2008) Surface characterization and direct electrochemistry of redox copper centers of bilirubin oxidase from fungi Myrothecium verrucaria. Bioelectrochemistry 74(1):101–110
9. McEvoy JP, Foord JS (2005) Direct electrochemistry of blue copper proteins at boron-doped diamond electrodes. Electrochim Acta 50(14):2933–2941
10. Sultana N, Schenkman JB, Rusling JF (2007) Direct electrochemistry of cytochrome P450 reductases in surfactant and polyion films. Electroanal 19(24):2499–2506
11. Zhao J, Zhu XL, Li T, Li GX (2008) Self-assembled multilayer of gold nanoparticles for amplified electrochemical detection of cytochrome c. Analyst 133(9):1242–1245
12. Ye T, Kaur R, Senguen FT, Michel LV, Bren KL, Elliott SJ (2008) Methionine ligand lability of type I Cytochromes c: detection of ligand loss using protein film voltammetry. J Am Chem Soc 130(21):6682–6683
13. Guo ZY, Zhang HN, Gai PP, Duan J (2011) Direct electrochemistry of cytochrome c entrapped in agarose hydrogel by protein film voltammetry. Russ J Electrochem 47(2):175–180
14. Elliott SJ, Bradley AL, Arciero DM, Hooper AB (2007) Protonation and inhibition of Nitrosomonas europaea cytochrome c peroxidase observed with protein film voltammetry. J Inorg Biochem 101(1):173–179
15. Sun W, Zhai ZQ, Wang DD, Liu SF, Jiao K (2009) Electrochemistry of hemoglobin entrapped in a Nafion/nano-ZnO film on carbon ionic liquid electrode. Bioelectrochemistry 74(2):295–300
16. Wen YL, Yang XD, Hu GH, Chen SH, Jia NQ (2008) Direct electrochemistry and biocatalytic activity of hemoglobin entrapped into gellan gum and room temperature ionic liquid composite system. Electrochim Acta 54(2):744–748
17. Long JS, Silvester DS, Wildgoose GG, Surkus AE, Flechsig GU, Compton RG (2008) Direct electrochemistry of horseradish peroxidase immobilized in a chitosan-[C(4)mim][BF(4)] film: determination of electrode kinetic parameters. Bioelectrochemistry 74(1):183–187
18. Zhang Y, He PL, Hu NF (2004) Horseradish peroxidase immobilized in TiO_2 nanoparticle films on pyrolytic graphite electrodes: direct electrochemistry and bioelectrocatalysis. Electrochim Acta 49(12):1981–1988
19. Hoeben FJM, Meijer FS, Dekker C, Albracht SPJ, Heering HA, Lemay SG (2008) Toward single-enzyme molecule electrochemistry: [NiFe]-hydrogenase protein film voltammetry at nanoelectrodes. ACS Nano 2(12):2497–2504
20. Xu ZA, Gao N, Chen HJ, Dong SJ (2005) Biopolymer and carbon nanotubes interface prepared by self-assembly for studying the electrochemistry of microperoxidase-11. Langmuir 21(23):10808–10813

21. Liu Y, Wang MK, Zhao F, Guo ZH, Chen HJ, Dong SJ (2005) Direct electron transfer and electrocatalysis of microperoxidase immobilized on nanohybrid film. J Electroanal Chem 581(1):1–10

22. Cao W, Wei CM, Hu JB, Li QL (2008) Direct electrochemistry and electrocatalysis of myoglobin immobilized on gold nanoparticles/carbon nanotubes nanohybrid film. Electroanal 20(17):1925–1931

23. Kumar SA, Chen SM (2007) Myoglobin/arylhydroxylamine film modified electrode: direct electrochemistry and electrochemical catalysis. Talanta 72(2):831–838

24. Guo XH, Zhang H, Hu NF (2008) Myoglobin-loaded layer-by-layer films containing SiO(2) nanoparticles studied using electrochemistry. Nanotechnology 19(5):055709

25. Wijma HJ, Jeuken LJC, Verbeet MP, Armstrong FA, Canters GW (2007) Protein film voltammetry of copper-containing nitrite reductase reveals reversible inactivation. J Am Chem Soc 129(27):8557–8565

26. Shen M, Wang J, Yang M, Li GX (2011) Direct electrochemistry of the Ti(IV)-transferrin complex: probing into the transport of Ti(IV) by human serum transferrin. Electrochem Commun 13(2):114–116

27. Aubin-Tam ME, Hamad-Schifferli K (2005) Gold nanoparticle cytochrome c complexes: the effect of nanoparticle ligand charge on protein structure. Langmuir 21(26):12080–12084

28. Ciampi S, Gooding JJ (2010) Direct electrochemistry of cytochrome c at modified Si(100) electrodes. Chem-Eur J 16(20):5961–5968

29. Eddowes MJ, Hill HAO (1979) Electrochemistry of horse heart cytochrome-c. J Am Chem Soc 101(16):4461–4464

30. Antonini E (1965) Interrelationship between structure and function in hemoglobin and myoglobin. Physiol Rev 45(1):123–170

31. Linberg R, Conover CD, Shum KL, Shorr RGL (1998) Hemoglobin based oxygen carriers: how much methemoglobin is too much? Artif Cell Blood Sub 26(2):133–148

32. Hultquist DE, Sannes LJ, Juckett DA (1984) Catalysis of methemoglobin reduction. Curr Top Cell Regul 24:287–300

33. Zheng N, Zhou X, Yang WY, Li XJ, Yuan ZB (2009) Direct electrochemistry and electrocatalysis of hemoglobin immobilized in a magnetic nanoparticles-chitosan film. Talanta 79(3):780–786

34. Zhou H, Gan X, Liu T, Yang QL, Li GX (2005) Effect of nano cadmium sulfide on the electron transfer reactivity and peroxidase activity of hemoglobin. J Biochem Bioph Meth 64(1):38–45

35. Yin F, Shin HK, Kwon YS (2005) Direct electrochemistry of hemoglobin immobilized on gold electrode by Langmuir-Blodgett technique. Biosens Bioelectron 21(1):21–29

36. Fan CH, Li GX, Zhu JQ, Zhu DX (2000) A reagentless nitric oxide biosensor based on hemoglobin-DNA films. Anal Chim Acta 423(1):95–100

37. Ma X, Liu XJ, Xiao H, Li GX (2005) Direct electrochemistry and electrocatalysis of hemoglobin in poly-3-hydroxybutyrate membrane. Biosens Bioelectron 20(9):1836–1842

38. Liu XJ, Xiao H, Shang LB, Wang XY, Li GX (2005) Electrochemical studies of hemoglobin and myoglobin embedded in dipalmitoylphosphatidic acid films. Anal Lett 38(3):453–462

39. Zhou H, Yang RW, Xu Y, Han K, Li GX (2005) Direct electrochemistry and catalytic activity of hemoglobin and myoglobin entrapped in PEG film. Anal Lett 38(13):2103–2115

40. Zhou H, Chen Z, Yang RW, Shang LB, Li GX (2006) Direct electrochemistry and electrocatalysis of hemoglobin in lactobionic acid film. J Chem Technol Biot 81(1):58–61

41. Liu SQ, Lin BP, Yang XD, Zhang QQ (2007) Carbon-nanotube-enhanced direct electron-transfer reactivity of hemoglobin immobilized on polyurethane elastomer film. J Phys Chem B 111(5):1182–1188

42. Zheng W, Li J, Zheng YF (2008) An amperometric biosensor based on hemoglobin immobilized in poly(epsilon-caprolactone) film and its application. Biosens Bioelectron 23(10):1562–1566

43. Yu JJ, Ma JR, Zhao FQ, Zeng BZ (2007) Direct electron-transfer and electrochemical catalysis of hemoglobin immobilized on mesoporous Al_2O_3. Electrochim Acta 53(4):1995–2001

44. Sun JY, Huang KJ, Zhao SF, Fan Y, Wu ZW (2011) Direct electrochemistry and electro-catalysis of hemoglobin on chitosan-room temperature ionic liquid-TiO(2)-graphene nano-composite film modified electrode. Bioelectrochemistry 82(2):125–130

45. Rosales-Rivera LC, Acero-Sanchez JL, Lozano-Sanchez P, Katakis I, O'Sullivan CK (2011) Electrochemical immunosensor detection of antigliadin antibodies from real human serum. Biosens Bioelectron 26(11):4471–4476

46. Yeo J, Park JY, Bae WJ, Lee YS, Kim BH, Cho YJ, Park SM (2009) Label-free electrochem-ical detection of the p53 core domain protein on its antibody immobilized electrode. Anal Chem 81(12):4770–4777

47. Yang HC, Yuan R, Chai YQ, Mao L, Su HL, Jiang W, Liang M (2011) Electrochemical immunosensor for detecting carcinoembryonic antigen using hollow Pt nanospheres-labeled multiple enzyme-linked antibodies as labels for signal amplification. Biochem Eng J 56(3):116–124

48. Wang J, Cao Y, Xu YY, Li GX (2009) Colorimetric multiplexed immunoassay for sequential detection of tumor markers. Biosens Bioelectron 25(2):532–536

49. Liu HY, Malhotra R, Peczuh MW, Rusling JF (2010) Electrochemical immunosensors for antibodies to peanut allergen ara h2 using gold nanoparticle-peptide films. Anal Chem 82(13):5865–5871

50. Munoz LE, Gaipl US, Herrmann M (2008) Predictive value of anti-dsDNA autoantibodies: importance of the assay. Autoimmun Rev 7(8):594–597

51. Ricci F, Adornetto G, Moscone D, Plaxco KW, Palleschi G (2010) Quantitative, reagentless, single-step electrochemical detection of anti-DNA antibodies directly in blood serum. Chem Commun 46(10):1742–1744

52. White RJ, Kallewaard HM, Hsieh W, Patterson AS, Kasehagen JB, Cash KJ, Uzawa T, Soh HT, Plaxco KW (2012) Wash-free, electrochemical platform for the quantitative, multi-plexed detection of specific antibodies. Anal Chem 84(2):1098–1103

53. Chen AQ, Du D, Lin YH (2012) Highly sensitive and selective immuno-capture/electro-chemical assay of acetylcholinesterase activity in red blood cells: a biomarker of exposure to organophosphorus pesticides and nerve agents. Environ Sci Technol 46(3):1828–1833

54. Ellington AD, Szostak JW (1990) Invitro selection of Rna molecules that bind specific ligands. Nature 346(6287):818–822

55. Tuerk C, Gold L (1990) Systematic evolution of ligands by exponential enrichment—Rna ligands to bacteriophage-T4 DNA-polymerase. Science 249(4968):505–510

56. Irvine D, Tuerk C, Gold L (1991) Selexion—systematic evolution of ligands by exponential enrichment with integrated optimization by nonlinear-analysis. J Mol Biol 222(3):739–761

57. Dollins CM, Nair S, Sullenger BA (2008) Aptamers in immunotherapy. Hum Gene Ther 19(5):443–450

58. Kai Z, Zhu XL, Jing W, Xu LL, Li GX (2010) Strategy to fabricate an electrochemi-cal aptasensor: application to the assay of adenosine deaminase activity. Anal Chem 82(8):3207–3211

59. Song SP, Wang LH, Li J, Zhao JL, Fan CH (2008) Aptamer-based biosensors. Trac-Trend Anal Chem 27(2):108–117

60. Lee JF, Stovall GM, Ellington AD (2006) Aptamer therapeutics advance. Curr Opin Chem Biol 10(3):282–289

61. Bruno JG, Carrillo MR, Cadieux CL, Lenz DE, Cerasoli DM, Phillips T (2009) DNA aptamers developed against a soman derivative cross-react with the methylphosphonic acid core but not with flanking hydrophobic groups. J Mol Recognit 22(3):197–204

62. Jenison RD, Gill SC, Pardi A, Polisky B (1994) High-resolution molecular discrimination by Rna. Science 263(5152):1425–1429

63. Liu XM, Zhang DJ, Cao GJ, Yang G, Ding HM, Liu G, Fan M, Shen BF, Shao NS (2003) RNA aptamers specific for bovine thrombin. J Mol Recognit 16(1):23–27

64. Ke YG, Lindsay S, Chang Y, Liu Y, Yan H (2008) Self-assembled water-soluble nucleic acid probe tiles for label-free RNA hybridization assays. Science 319(5860):180–183

65. Li Y, Lee HJ, Corn RM (2007) Detection of protein biomarkers using RNA aptamer microarrays and enzymatically amplified surface plasmon resonance imaging. Anal Chem 79(3):1082–1088

66. Swensen JS, Xiao Y, Ferguson BS, Lubin AA, Lai RY, Heeger AJ, Plaxco KW, Soh HT (2009) Continuous, real-time monitoring of cocaine in undiluted blood serum via a micro-fluidic, electrochemical aptamer-based sensor. J Am Chem Soc 131(12):4262–4266

67. Ferreira CSM, Bisland S, Gariepy J (2007) Novel strategy for targeted photodynamic therapy of breast cancer using a chlorin e6 conjugated-aptamer. Mol Cancer Ther 6(12):3558s–3559s

68. Rusconi CP, Scardino E, Layzer J, Pitoc GA, Ortel TL, Monroe D, Sullenger BA (2002) RNA aptamers as reversible antagonists of coagulation factor IXa. Nature 419(6902):90–94

69. Savran CA, Knudsen SM, Ellington AD, Manalis SR (2004) Micromechanical detection of proteins using aptamer-based receptor molecules. Anal Chem 76(11):3194–3198

70. Freitas SC, Barbosa MA, Martins MCL (2010) The effect of immobilization of throm-bin inhibitors onto self-assembled monolayers on the adsorption and activity of thrombin. Biomaterials 31(14):3772–3780

71. Wang J, Cao Y, Chen GF, Li GX (2009) Regulation of thrombin activity with a bifunctional aptamer and hemin: development of a new anticoagulant and antidote pair. ChemBioChem 10(13):2171–2176

72. Radi AE, Sanchez JLA, Baldrich E, O'Sullivan CK (2005) Reusable impedimetric aptasen-sor. Anal Chem 77(19):6320–6323

73. Deng CY, Chen JH, Nie Z, Wang MD, Chu XC, Chen XL, Xiao XL, Lei CY, Yao SZ (2009) Impedimetric aptasensor with femtomolar sensitivity based on the enlargement of surface-charged gold nanoparticles. Anal Chem 81(2):739–745

74. Radi AE, Sanchez JLA, Baldrich E, O'Sullivan CK (2006) Reagentless, reusable, ultrasensi-tive electrochemical molecular beacon aptasensor. J Am Chem Soc 128(1):117–124

75. Ding CF, Ge Y, Lin JM (2010) Aptamer based electrochemical assay for the determina-tion of thrombin by using the amplification of the nanoparticles. Biosens Bioelectron 25(6):1290–1294

76. Xiao Y, Lubin AA, Heeger AJ, Plaxco KW (2005) Label-free electronic detection of thrombin in blood serum by using an aptamer-based sensor. Angew Chem Int Edit 44(34):5456–5459

77. Hansen JA, Wang J, Kawde AN, Xiang Y, Gothelf KV, Collins G (2006) Quantum-dot/aptamer-based ultrasensitive multi-analyte electrochemical biosensor. J Am Chem Soc 128(7):2228–2229

78. Wang XY, Dong P, Yun W, Xu Y, He PG, Fang YZ (2009) A solid-state electrochemilumi-nescence biosensing switch for detection of thrombin based on ferrocene-labeled molecular beacon aptamer. Biosens Bioelectron 24(11):3288–3292

79. Cho H, Baker BR, Wachsmann-Hogiu S, Pagba CV, Laurence TA, Lane SM, Lee LP, Tok JBH (2008) Aptamer-based SERRS sensor for thrombin detection. Nano Lett 8(12):4386–4390

80. Hianik T, Grman I, Karpisova I (2009) The effect of DNA aptamer configuration on the sen-sitivity of detection thrombin at surface by acoustic method. Chem Commun 41:6303–6305

81. Wang YY, Liu B (2009) Conjugated polyelectrolyte-sensitized fluorescent detection of thrombin in blood serum using aptamer-immobilized silica nanoparticles as the platform. Langmuir 25(21):12787–12793

82. Pinto A, Redondo MCB, Ozalp VC, O'Sullivan CK (2009) Real-time apta-PCR for 20,000-fold improvement in detection limit. Mol BioSyst 5(5):548–553

83. Xu DK, Xu DW, Yu XB, Liu ZH, He W, Ma ZQ (2005) Label-free electrochemical detec-tion for aptamer-based array electrodes. Anal Chem 77(16):5107–5113

84. Wang J, Meng WY, Zheng XF, Liu SL, Li GX (2009) Combination of aptamer with gold nanoparticles for electrochemical signal amplification: application to sensitive detection of platelet-derived growth factor. Biosens Bioelectron 24(6):1598–1602

85. Roger PP, Christophe D, Dumont JE, Pirson I (1997) The dog thyroid primary culture sys-tem: a model of the regulation of function, growth and differentiation expression by cAMP and other well-defined signaling cascades. Eur J Endocrinol 137(6):579–598

86. Vandenberghe P, Freeman GJ, Nadler LM, Fletcher MC, Kamoun M, Turka LA, Ledbetter JA, Thompson CB, June CH (1992) Antibody and B7/Bb1-mediated ligation of the Cd28 receptor induces tyrosine phosphorylation in human T-Cells. J Exp Med 175(4):951–960

87. Yang Y, Guo LH, Qu N, Wei MY, Zhao LX, Wan B (2011) Label-free electrochemical measurement of protein tyrosine kinase activity and inhibition based on electro-catalyzed tyrosine signaling. Biosens Bioelectron 28(1):284–290

88. Miao P, Ning LM, Li XX, Li PF, Li GX (2012) Electrochemical strategy for sensing protein phosphorylation. Bioconjugate Chem 23(1):141–145

89. Wang J, Shen M, Cao Y, Li GX (2010) Switchable "On–Off" electrochemical technique for detection of phosphorylation. Biosens Bioelectron 26(2):638–642

90. Miao P, Ning LM, Li XX, Shu YQ, Li GX (2011) An electrochemical alkaline phosphatase biosensor fabricated with two DNA probes coupled with lambda exonuclease. Biosens Bioelectron 27(1):178–182

91. Gan X, Liu T, Zhong J, Liu XJ, Li GX (2004) Effect of silver nanoparticles on the electron transfer reactivity and the catalytic activity of myoglobin. ChemBioChem 5(12):1686–1691

92. Huang YX, Zhang WJ, Xiao H, Li GX (2005) An electrochemical investigation of glucose oxidase at a US nanoparticles modified electrode. Biosens Bioelectron 21(5):817–821

93. Liu T, Zhong J, Gan X, Fan CH, Li GX, Matsuda N (2003) Wiring electrons of cytochrome c with silver nanoparticles in layered films. ChemPhysChem 4(12):1364–1366

94. Wang J, Cao Y, Li Y, Liang ZQ, Li GX (2011) Electrochemical strategy for detection of phosphorylation based on enzyme-linked electrocatalysis. J Electroanal Chem 656(1–2):274–278

95. Kim SN, Rusling JF, Papadimitrakopoulos F (2007) Carbon nanotubes for electronic and electrochemical detection of biomolecules. Adv Mater 19(20):3214–3228

96. Shao YY, Wang J, Wu H, Liu J, Aksay IA, Lin YH (2010) Graphene based electrochemical sensors and biosensors: a review. Electroanal 22(10):1027–1036

97. Topoglidis E, Discher BM, Moser CC, Dutton PL, Durrant JR (2003) Functionalizing nanocrystalline metal oxide electrodes with robust synthetic redox proteins. ChemBioChem 4(12):1332–1339

98. McKenzie KJ, Marken F, Opallo M (2005) TiO$_2$ phytate films as hosts and conduits for cytochrome c electrochemistry. Bioelectrochemistry 66(1–2):41–47

99. Zhou H, Liu L, Yin K, Liu SL, Li GX (2006) Electrochemical investigation on the catalytic ability of tyrosinase with the effect of nano titanium dioxide. Electrochem Commun 8(7):1168–1172

100. Miao P, Shen M, Ning LM, Chen GF, Yin YM (2011) Functionalization of platinum nanoparticles for electrochemical detection of nitrite. Anal Bioanal Chem 399(7):2407–2411

101. Higuchi A, Siao YD, Yang ST, Hsieh PV, Fukushima H, Chang Y, Ruaan RC, Chen WY (2008) Preparation of a DNA aptamer-Pt complex and its use in the colorimetric sensing of thrombin and anti-thrombin antibodies. Anal Chem 80(17):6580–6586

102. Cao Y, Wang J, Xu YY, Li GX (2010) Combination of enzyme catalysis and electrocatalysis for biosensor fabrication: application to assay the activity of indoleamine 2,3-dioxygensae. Biosens Bioelectron 26(1):87–91

103. Novoselov KS, Geim AK, Morozov SV, Jiang D, Zhang Y, Dubonos SV, Grigorieva IV, Firsov AA (2004) Electric field effect in atomically thin carbon films. Science 306(5696):666–669

104. Zhao J, Chen GF, Zhu L, Li GX (2011) Graphene quantum dots-based platform for the fabrication of electrochemical biosensors. Electrochem Commun 13(1):31–33

105. Wan Y, Deng WP, Su Y, Zhu XH, Peng C, Hu HY, Peng HZ, Song SP, Fan CH (2011) Carbon nanotube-based ultrasensitive multiplexing electrochemical immunosensor for cancer biomarkers. Biosens Bioelectron 30(1):93–99

Chapter 4
Electrochemical Analysis of Cells

Abstract The cell is the basic unit of life, which plays a key role in the development of organisms and participates in almost all physiological processes in vivo. Since the physiological activities of cells are often related to electron transfer and/or electroactive species, electrochemistry is proven to be an effective technique for the analysis of cells, which can be further used in disease diagnosis and drug screening. Therefore, electrochemical analysis of cells has attracted a great many research interests. In recent years, with the development of surface modification technology, molecular recognition and nanotechnology, more and more electrodes with high biocompatibility can be used for cell immobilization, which has greatly promoted the electrochemical analysis of cells.

Keywords Cell • Electrochemical analysis • Cell membrane • Electrochemical cytosensor • Intracellular analysis • Cell apoptosis • Cancer cell detection • Nanomaterials-based amplification

The cell is the basic functional unit of life with genetic totipotency. It is also the basis for organism growth and development with orderly self-controlled metabolic system. A cell is sophisticated combination of matter, energy and information with highly ordered self-assembly and self-organizing capacity. Cells always play key roles in the developmental and physiological processes of organisms. Therefore, the understanding of cells is the basis of biologic research, and the analysis of cells is of great importance to all the fields of biologic sciences as well as biomedical applications.

Since the physiological activities of cells, such as metabolism, respiration, photosynthesis, transmembrane transport, etc., are usually coupled with the transfer of electron and/or related with electroactive species, electroanalytical methods have become useful tools for the analysis of cells. Meanwhile, since abnormal cell activities may lead to a lot of diseases, electrochemical analysis of cells is also proven to be effective in disease diagnosis and drug screening.

On the other hand, electrochemistry is a powerful technique to study electron transfer process, which offers many advantages including simplicity, highly sensitivity, convenience and low cost. So, electrochemical bioanalysis

G. Li and P. Miao, *Electrochemical Analysis of Proteins and Cells*, SpringerBriefs   43
in Molecular Science, DOI: 10.1007/978-3-642-34252-3_4, © The Author(s) 2013

has been developed very rapidly in recent years [1–7]. Meanwhile, coupled with some techniques such as surface modification, nanotechnology, molecular recognition, etc., various CMEs have been prepared to make the electrochemical analysis of cells more achievable. Therefore, the inner sub-cellular response mechanisms can be better understood, and the related physiological activities can be more thoroughly explored. Moreover, since the enzymes extracted from cells can often maintain their biologic activities and the studies on these enzymes with electrochemical technique have provided a lot of research methods as well as valuable information on the enzymes, electrochemical analysis of cells has made it more effective to understand in-depth the law of motion of the organisms at the cellular level. Besides, electrochemical analysis of cells is also helpful for drug design, drug side effects controlling, microorganisms monitoring and the other related scientific theoretical and experimental basis. So, cell-based electrochemical analysis has been widely used in the fields of pharmacology, medicine, cell biology, toxicology, neuroscience and environmental monitoring [8].

In this chapter, we mainly summarize the progress in electrochemical analysis of cell membrane and the inner part of cells, the electrochemical detection of cell apoptosis and the identification and quantification of cancer cells. In addition, the applications of nanomaterials in the electrochemical analysis of cells have also been discussed.

4.1 Electrochemical Analysis of Cell Membrane and the Inner Part

Compared with proteins, cells are more complicated units with elaborate subcellular machinery for biologic events [9, 10]. Meanwhile, living cells are electrochemical dynamic systems, which make it more difficult to obtain the electrochemical information. Nevertheless, the electron generation and transfer in a cell, especially the electron transfer across the cell membrane, can indeed indicate the status of cells in life process [11]. As is well known, cell membrane separates the interior of a cell from the environment. It mainly consists of phospholipids bilayer. The transportation of substance inside or outside a cell is controlled by the selectively permeable cell membrane. Therefore, the penetrability of cell membrane has been firstly studied. For instance, Huang et al. [12] have fabricated a phosphatidylcholine bilayer membrane on the surface of a PGE to mimic cell membrane. They use two electrochemical probes, $Fe(CN)_6^{3-/4-}$ and $Ru(bpy)_3^{2+}$, to separately measure the electrochemical response of the ion-channel behavior of annexin V. As shown in Fig. 4.1, in the presence of annexin V, the electrochemical responses of both $Fe(CN)_6^{3-/4-}$ and $Ru(bpy)_3^{2+}$ can be enhanced, indicating the formation of ion-permeable pores in the membrane. Further studies reveal that this procedure can be affected by pH value and the concentration of Ca^{2+}.

Fig. 4.1   a Cyclic
voltammograms of
Fe(CN)$_6^{3-/4-}$ solution
obtained at bare
PGE (*dashed*), the
annexin V-embedded
phosphatidylcholine bilayer
membranes (*BLM*)-modified
electrode (*solid*) and the
BLM alone modified
electrode (*dotted*);
  b Cyclic voltammograms of
Ru(bpy)$_3^{2+}$ solution obtained
at the bare PGE (*solid*),
the annexin V-embedded
BLM-modified electrode
(*dotted*) and the BLM alone
modified electrode (*dashed*).
(Reprinted from Ref. [12],
Copyright 2008, with
permission from Elsevier)

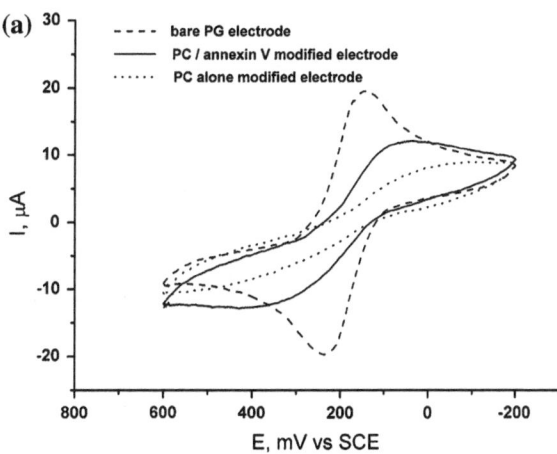

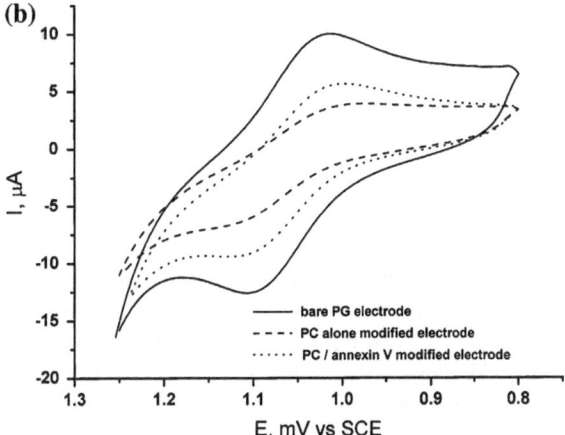

Some research groups have tried to immobilize some protein complex such as photosystem I and II onto electrode surface to mimic the electron transfer in vitro [13, 14]. For instance, Alcantara et al. have obtained direct, reversible peaks of the spinach photosystem II reaction center in a lipid film–modified electrode. They have also analyzed the three pairs of reduction–oxidation peaks, which are assigned to oxygen-evolving complex tetramanganese cluster, quinone and pheophytin, respectively.

Cell surface glycoproteins are integral membrane proteins that contain oligosaccharide chains (glycans). These proteins play a key role in cell–cell interactions. So, Shao et al. [15] have developed a biocompatible poly(diallyldimethylammonium)-doped poly(dimethylsiloxane) film to electrochemically measure the cell surface glycoprotein. The interface they construct can retain the viability of the immobilized cells. They then use p-glycoprotein antibody and its secondary antibody

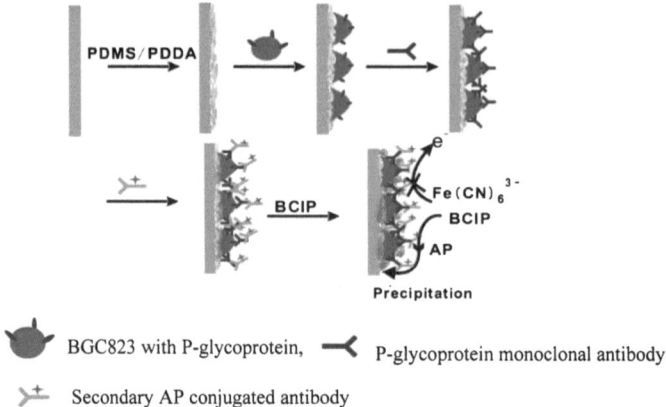

Fig. 4.2 Schematic representation of the designed sensor for the evaluation of cell surface gly-coprotein. *AP* alkaline phosphatase. (Reprinted with the permission from Ref. [15]. Copyright 2009 American Chemical Society)

labeled with alkaline phosphatase to detect p-glycoproteins on cell surface, the strategy of which can be illustrated by Fig. 4.2.

Up to now, various membrane materials with excellent biocompatibility have been employed to immobilize cells on electrode surface for the analysis with electrochemical techniques [16, 17]. However, the studies are usually focused on the cell membrane, which cannot explore the inner part of the cells. So, Meng et al. [18] have proposed a universal method to fix cells on solid interface based on transfection techniques and further explored the inner part of the cells. Figure 4.3 shows the transfection process to immobilize cells. Firstly, exogenous thiolated DNA is modified on a gold electrode via gold–sulfur chemistry. Then, the DNA molecules are transfected into the cells due to the presence of calcium ions. This procedure may also drag the 293T cells onto the surface of the electrode. In fact, the DNA molecules can not only fix the cells on the electrode surface, but also act as wires of electron transfer. Therefore, the electron transfer between the electroactive species inside the cells and the electrode is achieved, and the electrochemical signals of both methylene blue and kaempferol, which have been pre-cultured with the cells, can be obtained. Moreover, the obtained electrochemical response will increase with the concentration of the electrochemical species inside the cells. So, this method makes it possible to study the inner part of cells by using electrochemical technique.

Based on the above transfection strategy, Liu et al. [19] have proposed a method for the detection of cells by using ferrocene as the electrochemical reporter. Firstly, they design a single-strand DNA probe dually labeled with 3′-SH and 5′-ferrocene. Then, the DNA probe is immobilized onto the surface of a gold electrode. Since the ferrocene moiety may easily approach the electrode due to the nice flexibility of the designed single-strand DNA, a large faradaic current can be yielded. Nevertheless, after the transfection of the DNA probe into the detected cells, the faradaic current decreases sharply with the increase in cells (Fig. 4.4a), since the insertion of DNA probe into the cells will keep ferrocene away from the

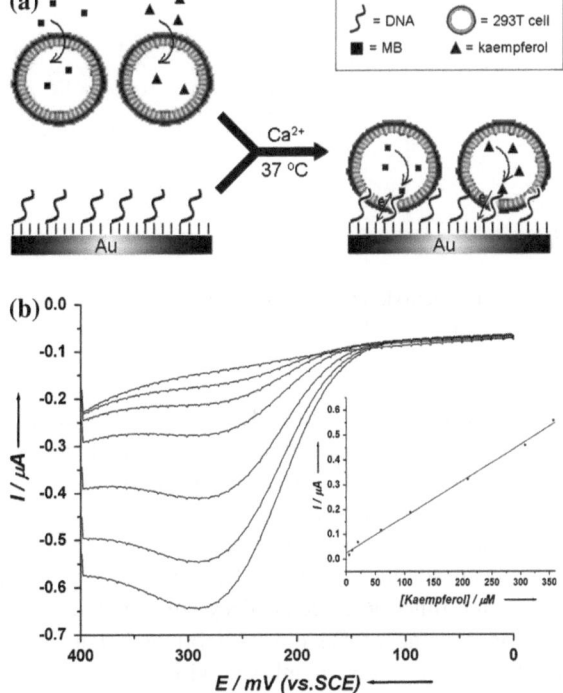

Fig. 4.3 **a** Schematic illustration of the immobilization of cells on a gold electrode surface and the execution of the electric communication between electroactive species inside the cells and the electrode via the transfected surface-immobilized DNA; **b** Linear sweep voltammograms of PBS obtained at a DNA-modified gold electrode after the incubation with 293T cells, which have been previously treated with kaempferol of various concentrations. The *inset* shows the linear plot of the current peak versus the concentration of kaempferol. *MB* methylene blue. (Reprinted with the permission from Ref. [18]. Copyright 2009 American Chemical Society)

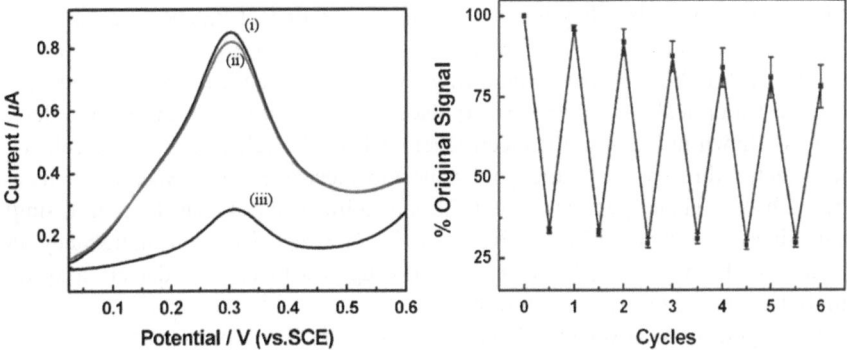

Fig. 4.4 **a** Differential pulse voltammograms obtained in PBS at the ferrocene-DNA-modified electrode before (*i*) and after reaction with SMMC-7721 cells (10^5 cells/mL) in the absence (*ii*) and presence (*iii*) of Ca^{2+}; **b** Signal conversion and retrieve (on–off–on) during 6 times of regeneration by treating the electrode with lysis buffer. (Reproduced from Ref. [19] by permission of The Royal Society of Chemistry)

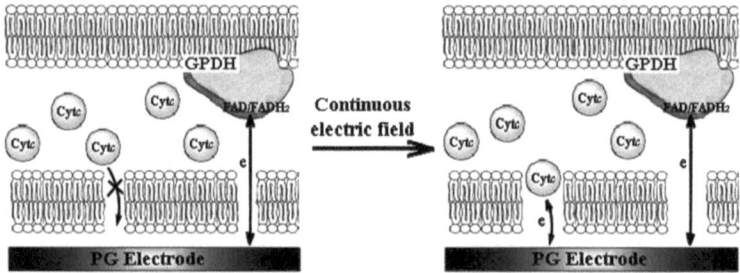

Fig. 4.5 Schematics of the electrode immobilization and reactions of mitochondrion. (Reproduced from Ref. [21] by permission of John Wiley & Sons Ltd)

electrode surface. The experiment results also confirm that the peak current will remain unchanged if Ca^{2+} is not involved in the transfection process, since Ca^{2+} is the necessity for the transfection. This electrochemical system can also be regenerated with the treatment of lysis buffer (Fig. 4.4b).

Besides cell membrane, electrochemical analysis of some macro objects like organelles has also been conducted. Organelles are the basic structures of cells which ensure the normal function of cells. They are usually enclosed within lipid bilayer. Mitochondria are one of the most important organelles, which provide energy for cell activities, so study of mitochondria has received great attention [20]. Zhao et al. [21] have examined the electrochemical behavior of mitochondria by co-immobilizing freshly extracted mitochondria, BSA and glutaraldehyde on the surface PGE (Fig. 4.5). Owing to the change of permeability of the outer mitochondrial membrane by the applied electric field, two pairs of redox peaks can be observed, assigned to the electron transfer reactions of cyt c and $FAD/FADH_2$, respectively. The redox wave of NADH can also be obtained when the membrane of mitochondria is destroyed.

The shape and function of cells are influenced or regulated by the biochemical processes of cells. Therefore, real-time observation of biochemical processes of cells is of great importance. Zheng et al. [22] have developed an electro-optical nanoprobe for the real-time monitoring of local biochemical processes in cells. The electro-optical nanoprobe can be placed on the cell membrane for the detection. As shown in Fig. 4.6, the electrochemical species released from the cells can be detected directly using the gold nanoring electrode. This system can measure both the oxidant generation and the intracellular antioxidant level in a single cell. Since the strength of oxidative stress is concerned with cell malignancy and a much high antioxidant level may cause drug resistance, this detection system might have broad utility in the future.

Cell cycle is a universal cell process involving the growth and proliferation of cells, which consists of many complex stages [23]. It is a series of events that take place in a cell, which lead to cell division and duplication. A cell in different phases may exhibit different behaviors even the environments are kept uniform. So, Kafi et al. [24] have proposed an electrochemical method for on-site

Fig. 4.6 Schematic illustration of the high-resolution optical detection of intracellular activities. (Reprinted from Ref. [22], Copyright 2011, with permission from Elsevier)

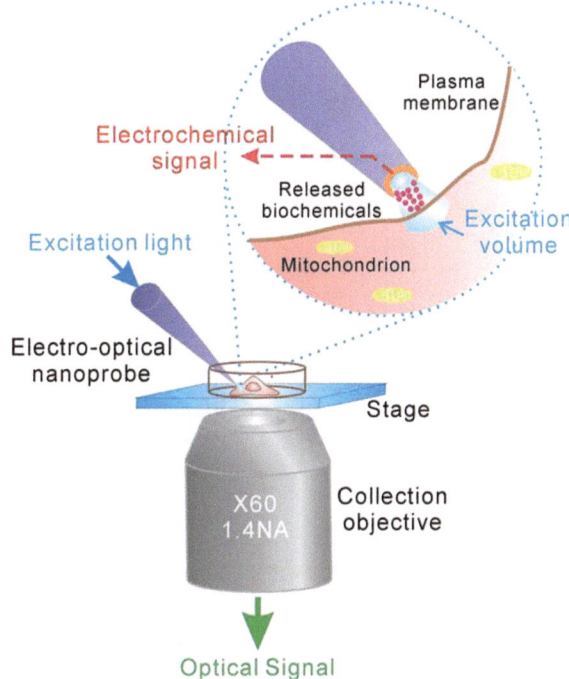

monitoring of cell cycle progression. As shown in Fig. 4.7, RGD-MAP-C peptide is designed for this study. The cysteine residue at the terminus of the peptide can be bound to the gold electrode via Au–S covalent bond and the RGD sequence may link integrin on cell membrane, which results in the attachment of cells on the electrode surface. Electron transfer at the cell–electrode interface then reflects the redox activity of the cells in different phases.

4.2   Electrochemical Analysis of Cell Apoptosis

Apoptosis is also termed as programmed cell death. It is a meticulous biochemical process involving many cell events such as cell turnover, the functioning of immune system, embryonic and soma development [25]. Apoptosis plays an essential role in the removal of unwanted or abnormal cells in the multicellular organisms [26]. The cell changes caused by apoptosis include blebbing, cell shrinkage, nuclear fragmentation, chromatin condensation and chromosomal DNA fragmentation. Different from necrosis, apoptosis produces cell fragments named as apoptotic bodies for phagocytic cells to remove before the contents of cells spill out and cause other damage.

Apoptosis has attracted much attention since the early 1990s, which not only exhibits as an important biologic phenomenon, but also implicates in an extensive

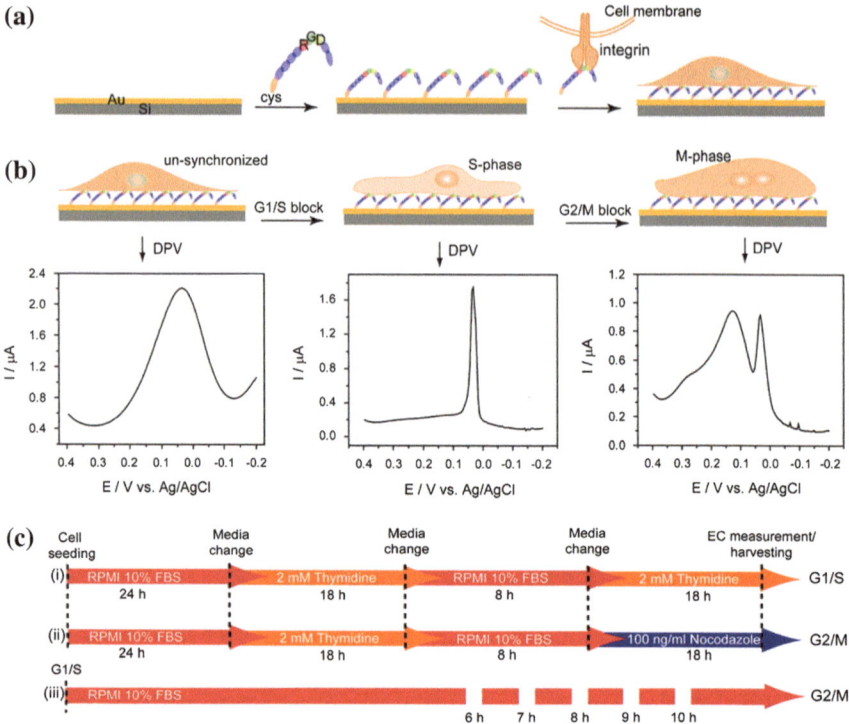

Fig. 4.7 Schematics of the electrochemical detection of cell cycle progression. **a** Fabrication of the RGD-MAP-C-based cell chip; **b** synchronized G1/S-phase (*middle*), G2/M-phase (*right*) and unsynchronized (*left*) cells with their respective DPV signals; **c** time course of cell treatment for synchronization in the (*i*) G1/S-phase and (*ii*) G2/M-phase and (*iii*) gradual progression of G1/S cells toward the G2/M-phase following time-dependent release from the G1/S block. (Reprinted with the permission from Ref. [24]. Copyright 2011 American Chemical Society)

variety of diseases. Apoptosis disorders include inappropriate activation or inhibition breaking the equilibrium between cell growth and cell death, which may directly related to many diseases, such as Alzheimer's disease, acquired immuno deficiency syndrome (AIDS), autoimmune diseases and all kinds of cancers [27]. Thus, in the pathophysiology, detection of apoptosis not only helps to improve the accuracy of the course of disease, but also can be used as the early signs for a therapeutic intervention effect judgment.

In early apoptotic cells, phosphatidylserine flips from the inside of the plasma membrane to the outside, and thus exposes to the extracellular surface. A Ca^{2+}-dependent phospholipid-binding protein, annexin V is usually used as a probe to specifically recognize the exposed phosphatidylserine on cell surface, indicating cell apoptosis [28]. Therefore, Tong et al. [29] have constructed a layer of AuNPs on an electrode surface for stable adsorption of large amount of annexin V. Subsequently, early apoptotic cells with exposed phosphatidylserine on the surface are bound to annexin V molecules that have been fixed on the electrode

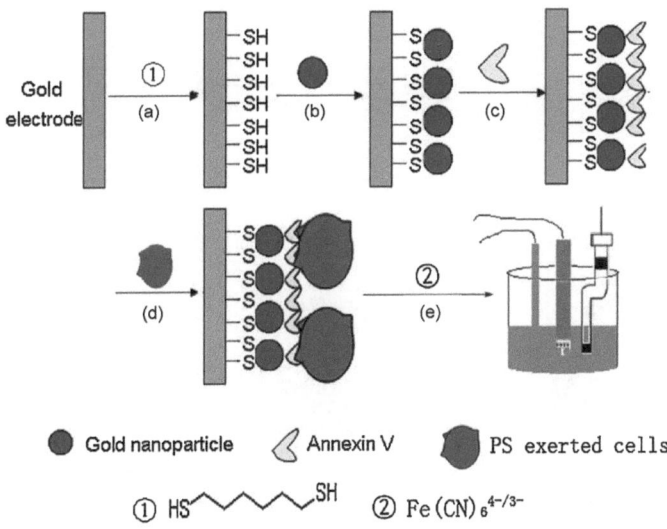

Fig. 4.8 Schematic representation of the formation of self-assembly layers **a** 1,6-hexanedithiol layer; **b** AuNPs layer; **c** Annexin V layer and the detection procedures; **d** incubation with cells; and **e** electrochemical impedance spectroscopy detection. (Reprinted from Ref. [29], Copyright 2009, with permission from Elsevier)

surface, hindering the electron transfer between the electrochemical species, $Fe(CN)_6^{3-/4-}$, in the solution and the electrode (Fig. 4.8). Since the process can be characterized by the increase in impedance, a biosensor is thus developed demonstrating the great potential for rapid detection of cell apoptosis. Meanwhile, Liu et al. [30] demonstrated that annexin V in polyethylenimine film can maintain its high affinity with phosphatidylserine translocated from the inner to the outer plasma membrane of the apoptotic cells. So, based on the interaction between annexin V and phosphatidylserine, voltammetric detection of apoptosis with electrochemical technique can be achieved by using annexin V and polyethylenimine co-modified PGE. Moreover, if Ca^{2+}, the required ion for the function of annexin V, is added into the test solution, the redox wave of the electrochemical probe, $[Ru(NH_3)_5Cl]^{2+}$, will be further changed; thus, a simple and convenient method to detect apoptosis is proposed.

It has been known that apoptotic cells will present some overexpressed cysteine-dependent aspartate-specific proteinases (caspases). These proteinases can be the important biomarkers of apoptosis in living cells and cell lysates. Caspase-3 is the most important apoptosis executor, which participates in a dual endogenous and exogenous apoptotic pathway. Caspase-3 is able to specifically hydrolyze the substrate peptide containing Asp-Glu-Val-Asp (DEVD) sequence [31]. So, Xiao et al. [32] have made use of this feature and proposed an electrochemical method for the detection of apoptosis by using the marker of caspase-3 activity. After the substrate peptide is assembled on a gold electrode surface with the help of cysteine, the electrochemical reporter, ferrocene, which is attached

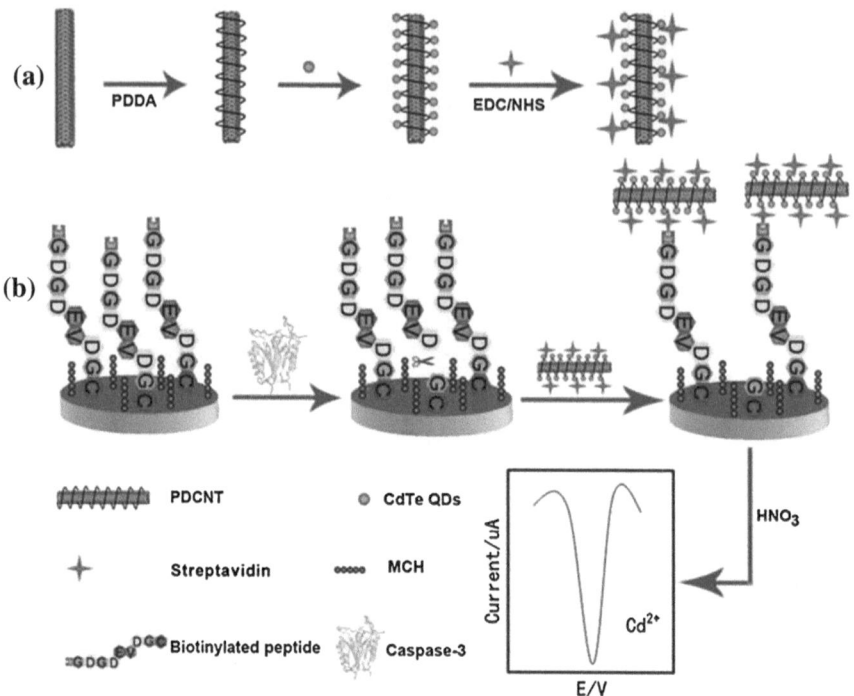

Fig. 4.9 a Scheme of the preparation of CNT-QDs-SA bioconjugates via layer-by-layer assembly; **b** Schematic representation of the electrochemical sensing strategy for caspase-3 activity. *QD* quantum dot; *SA* streptavidin. (Reproduced from Ref. [33] by permission of The Royal Society of Chemistry)

at the end of the peptide, may yield electrochemical signals. Nevertheless, since apoptotic cell lysate contains a large number of caspase-3, which will specifically recognize and cut the peptide substrate immobilized on the electrode surface, ferrocene group will be removed from the electrode. Therefore, the change in the electrochemical response can be used for the analysis of apoptosis.

In order to detect caspase-3 activity with a higher sensitivity, Zhang et al. [33] have utilized signal amplification strategy and layer-by-layer technology for such kind of study. In their work, quantum dots are used as electrochemical probes, while CNTs are labeled with streptavidin. As shown in Fig. 4.9, after the streptavidin-labeled CNTs are bound with the biotin-labeled peptides that have previously been immobilized on the surface of an electrode, anodic stripping voltammetry is applied to detect the stripping signal of cadmium. Therefore, signal amplification from CNTs can be achieved. Zhang et al. [34] have also conducted another work for ultrasensitive detection of apoptotic cells with electrochemical technique. In this work, they have fabricated a three-dimensional architecture via layer-by-layer strategy by combining nitrogen-doped CNTs and AuNPs, which can provide an effective matrix for annexin V immobilization with high stability

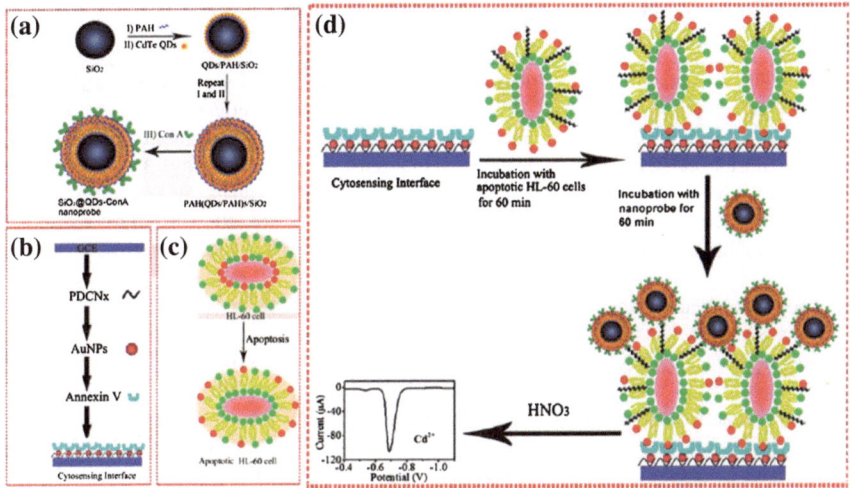

Fig. 4.10 Schematic illustration of **a** the preparation of SiO$_2$@QDs-Con A nanoprobe via layer-by-layer assembly; **b** the fabrication of the electrochemical cytosensing interface; **c** the exposure of phosphatidylserine residues on the outer surface of the cell membrane during apoptosis; **d** sandwich electrochemical sensing strategy for the apoptotic cells.*QD* quantum dot. (Reprinted with the permission from Ref. [34]. Copyright 2011 American Chemical Society)

and bioactivity (Fig. 4.10). Meanwhile, lectin-based nanoprobes are assembled on CdTe quantum dots–labeled silica nanospheres, which can specifically recognize carbohydrate on cell surface. Therefore, after a sandwich electrochemical sensing strategy is employed, detection of apoptotic cells can be achieved with a very high sensitivity.

4.3 Electrochemical Cytosensors

Cells have also been directly analyzed with electrochemical technique. To perform the direct electrochemical analysis, immobilization of cells onto the surface of an electrode is the key procedure. Therefore, many biocompatible matrices have been fabricated for cell immobilization, where the cells may well maintain their functions [35]. Among the prepared materials, biocompatible nanoparticles might be the best candidates, so they have been used for the construction of non-toxic biomimetic interface to immobilize cells [16, 36].

The adhesion of cells onto an electrode surface can be affected by many factors, such as topography, roughness, hydrophobicity, charges of the surface and specific proteins or nucleic acids [37]. Obviously, if the cell immobilization onto an electrode surface may mimic the natural cell adhesion on a solid support, the activity of cells will be better maintained.

Table 4.1 Characteristics of electrochemical analysis of cells[a] [8]

Modified electrode	Cells	Linear response range	Ref
Antibody/ITO IDAM	*Escherichia coli* O157:H7	4.36×10^5–4.36×10^8 cfu/mL	[40]
SAM/Au	*Saccharomyces cerevisiae*	10^2–10^8 cfu/mL	[41]
AuNPs-chitosan gel/GCE	K562 leukemia cells	1.34×10^4–1.34×10^8 cells/mL	[42]
Polyaniline/SPCE	K562 leukemia cells	10^4–10^7 cells/mL	[43]
Carbon nanofiber-chitosan/GCE	K562 leukemia cells	5×10^3–5×10^7 cells/mL	[44]
Antibody-biotin/neutravidin/ biotin-SAM/Au	*Escherichia coli*	10–10^3 cfu/mL	[45]
Antibody/epoxysilane/ITO	*Escherichia coli* O157:H7	6×10^4–6×10^7 cells/mL	[46]
Antibody magnetic NPs/Au IDAM	*Escherichia coli* O157:H7	7.4×10^4–7.4×10^7 cfu/mL	[47]
Antibody/Au IDAM	*Escherichia coli* O157:H7	10^4–10^7 cfu/mL	[48]

[a]Abbreviations: *IDAM* interdigitated array microelectrodes, *ITO* indium tin oxide, *NPs* nanoparticles, *SAM* self-assembly monolayer, *SPCE* screen printed carbon electrode

Cells always have excellent insulating properties, which may preserve the local ionic environment at electrode/solution interface [38–40]. Therefore, the changes in electrochemical impedance spectra can be used to monitor the cellular viability, number and adhesion status. On the other hand, the thickness of cell membranes usually lies in 5–10 nm, and the capacitance and resistance are in the ranges of 0.5–1.3 $\mu F\ cm^2$ and 10^2–10^5 $\Omega\ cm^2$, respectively. The cell attachment to an electrode surface can produce a barrier that hinders the electron transfer of redox probes with the electrode, which results in an increase in the electron transfer resistance (R_{et}). Therefore, electrochemical quantification of cells can be available by using electrochemical technique, since the increase in R_{et} is related to the amount of cells immobilized onto an electrode surface. For instance, Yang et al. [40] have conducted the study for the detection of *Escherichia coli* O157:H7 with a detection limit of 10^6 cfu/mL to be obtained. A number of other similar researches have also been carried out [41–48], and the detection of the cell lines and the corresponding detection ranges have been summarized in Table 4.1.

The measurement by electrochemical impedance spectroscopy can not only quantify the number of cells but also show the change of cells on the surface of an electrode [49]. As shown in Fig. 4.11, with the proliferation of cells on an electrode surface, the diameter of the semicircular part of spectra increases significantly, since the electron transfer resistance is gradually enhanced with the growth of the cells.

Cheng et al. [50] have developed another sensitive and selective membrane-based electrochemical nanobiosensor for quantitative label-free detection of *E. coli* cells and the analysis of cells viability. The innovation of this strategy lies in the blocking of nanochannels of a nanoporous alumina-membrane-modified electrode, while immune reaction occurs on the nanoporous membrane when

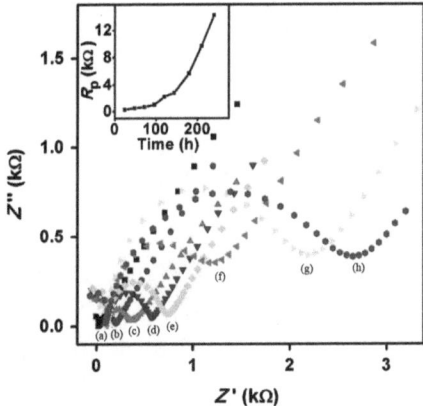

Fig. 4.11 Electrochemical impedance spectroscopic detection of cells adhesion at **a** bare and **b** zwi-film-modified electrodes and K562 cells proliferated on zwi-film-modified electrode after cell incubation for **c** 24, **d** 48, **e** 72, **f** 96, **g** 120 and **h** 140 h. Inset is the plot of the relationship between electron transfer resistance and proliferation time of K562 on zwi-film-modified electrodes. (Reprinted with the permission from Ref. [49]. Copyright 2005 American Chemical Society)

the *E.coli* cells exist (Fig. 4.12). Moreover, cyclic voltammetry can be applied to obtain direct electrochemical signal. This cell biosensor can obtain a low detection limit of 22 cfu/mL over a logarithmical linear range of $10–10^6$ cfu/mL.

4.4 Electrochemical Identification and Quantification of Cancer Cells

Cancer, also termed as malignant neoplasm, is a large group of diseases, in which cells divide and grow uncontrollably. It has one of the highest mortality among all diseases in the world. However, from the statistical work of World Health Organization, 30 % of cancer deaths can be avoided if early diagnosis and treatment are implemented [51, 52]. Therefore, early diagnosis of cancer can improve survival ratio of cancer patients. It is also one of the most effective protection approaches against the threat of cancer. So, electrochemical analysis of cancer cells has attracted great attention in recent years.

Telomeres and telomerase are closely related to the abnormal proliferation of tumor, and the expression of telomerase activity of human malignant tumors has been known to be abnormal, so analysis of telomerase activity has been an effective way for diagnosis and treatment for cancers [53, 54]. Shao et al. [55] have proposed an electrochemical method for sensitive assay of telomerase activity based on the strategy that telomerase extracted from cancer cells can trigger the extension of nucleic acid primers immobilized on an electrode surface.

Fig. 4.12 The fabrication of the nanoporous membrane-based biosensor for *Escherichia coli* analysis. (Reprinted with the permission from Ref. [50]. Copyright 2011 American Chemical Society)

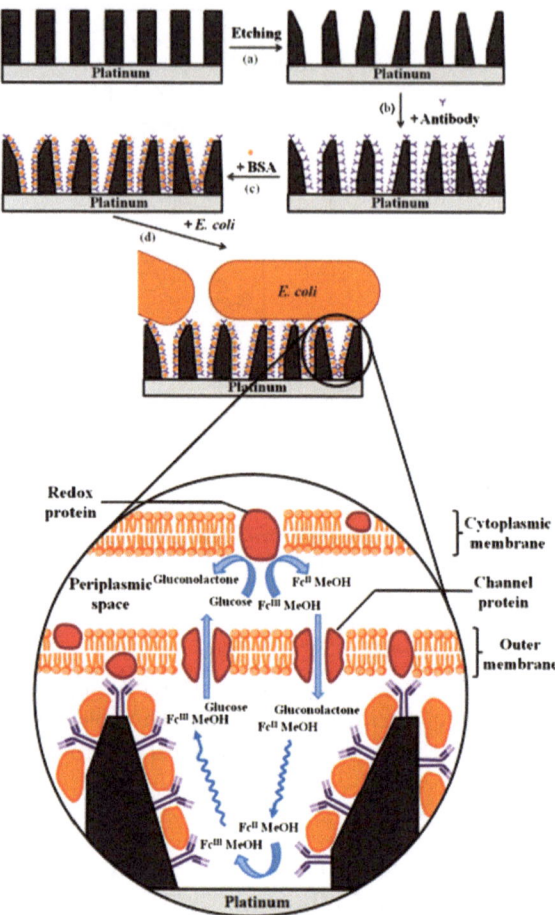

Consequently, a growing number of guanine-rich telomeric repeats can be added to the primers; thus, gradual increase in guanine oxidation signal can be obtained. This method can detect telomerase activity from 3,000 HeLa cells, so the sensitivity of this method can be even better than the traditional PCR-based methods.

Some research workers in this lab have also proposed a PCR-based strategy to sensitively identify cancer cells by using electrochemical technique based on the detection of telomerase activity in cell lysates [56]. After PCR amplification of telomerase extension product, DPV is employed to measure the electrochemical response of guanine; thus, sensitive quantification of cancer cells can be also achieved. Meanwhile, Yang et al. [57] have developed a label-free electrochemical impedance sensor for the assay of telomerase activity without using any amplification technique, and the sensitivity of this method can be as low as for the detection of 1,000 HeLa cells. Recently, Li et al. [58] further improved the detection limit to be 10 cells. The uses of AuNPs-mediated signal amplification strategy, instead

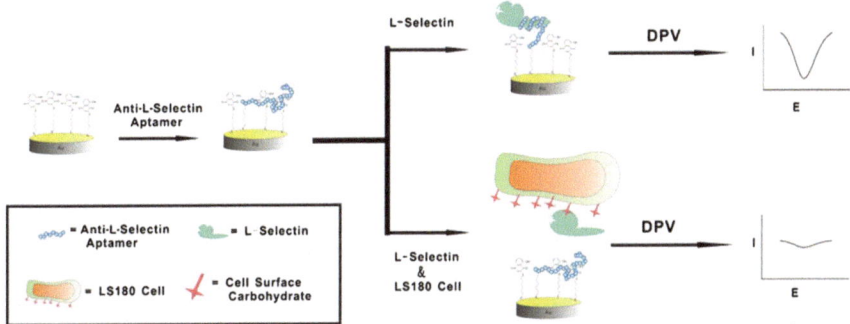

Fig. 4.13 Schematics of the electrochemical detection of cells and the analysis of the surface carbohydrates. (Reprinted from Ref. [64], with kind permission from Springer Science + Business Media)

of PCR amplification of telomerase extension product which is also laborious and time consuming, have promised the extremely high sensitivity.

In recent years, aptamers have been widely used for the fabrication of biosensors [59–61], so many aptamer-based electrochemical studies have been conducted for the detection of tumor protein biomarkers, such as human sticky protein-1 (MUC1), platelet-derived growth factor (PDGF), thrombin or even tumor cells, for instance human T-cell acute lymphoblastic leukemia cells CCRF-CEM [62, 63]. Shao et al. [64] have made use of anti-selectin aptamer for the analysis of tumor cells by successively immobilizing 5-hydroxy-3-hexanedithiol-1, 4-naphthoquinone (JUG$_{thio}$), the electrochemical reporter and anti-selectin aptamer on the surface of a gold electrode. As shown in Fig. 4.13, steric hindrance caused by the aptamer will reduce the electron transfer rate between JUG$_{thio}$ and the electrode; thus, only a weak electrochemical response can be obtained. However, after the combination of aptamer and L-selectin, the steric effect is eliminated; thus, a high electrochemical wave can be observed. Nevertheless, when human colon adenocarcinoma cells LS180 with abundant polysaccharides on the surface are introduced, the competitive binding of cells with the selectin molecules will make the aptamer to obstruct the electron transfer between the electrode and JUG$_{thio}$ again. Based on the relationship between the decreased peak current and the increased cell concentration, the method may report a linear detection range of 10^3–10^7 cells/mL for the detection of the cancer cells.

In many disease diagnoses, the identification of different cells plays a significant role, so some strategies have been proposed to identify different kind of cells. For instance, de la Escosura-Muniz et al. [65] have proposed an electrochemical method to distinguish target cells based on the immobilization of B-cell lymphoma HMy2 (HLA-DR positive) and prostate cancer PC-3 (HLA-DR negative) on the surface of a carbon-coated electrode, where the cells can grow well, similar to the state in plastic culture flasks. Through immune response between antihuman DR monoclonal antibody (αDR) and HLA-DR expressed on the cell surface, αDR antibody modified with AuNPs is linked to HLA-DR-positive cells HMy2

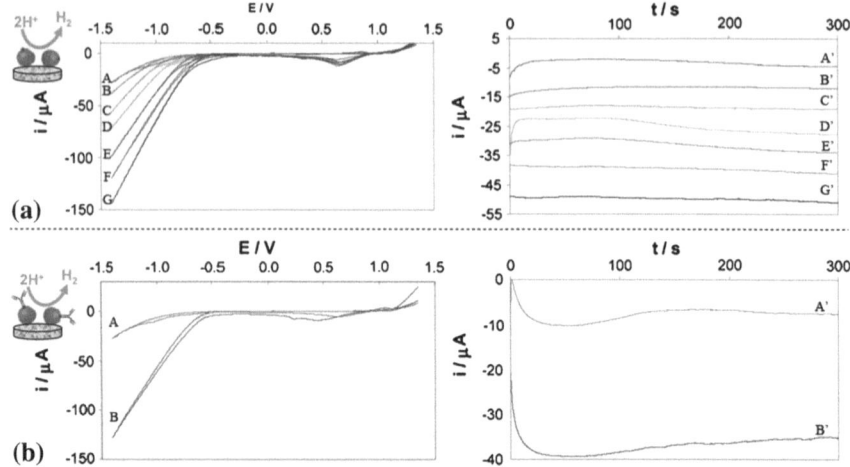

Fig. 4.14 a *Left* is the cyclic voltammograms for 1 M HCl solution (*blank curve, A*) and for increasing concentrations of AuNPs in 1 M HCl: (*B*) 0.96, (*C*) 4.8, (*D*) 24, (*E*) 120, (*F*) 600 and (*G*) 3,000 pM. Right is the chronoamperograms recorded by applying a potential of −1.00 V for 5 min, using a 1 M HCl solution (*blank curve, A′*) and the same AuNPs concentrations in 1 M HCl as detailed above (*B′–G′*); **b** *Left* is cyclic voltammograms recorded under the same conditions as in (*A*), which is for the blank (*curve A*) and for a solution of the conjugate AuNP/αDR (*curve B*). Right is chronoamperograms recorded under the same conditions as in **a**, which is for the blank (*A′*) and for the conjugate AuNPs/αDR (*B'*). Reprinted with the permission from Ref. [65]. Copyright 2009 American Chemical Society)

surface, while the reduction in hydrogen ion catalyzed by the nanoparticles can be achieved to produce electrochemical signals. Based on the catalytic signals, HLA-DR-positive HMy2 cells can be easily distinguished from the negative control cell line PC-3 (Fig. 4.14).

In order to more sensitively detect cancer cells, some signal amplification strategies have been proposed. For instance, Li et al. [66] have reported an approach by using electrochemical current rectifier for the detection of folate receptor–rich cancer cells. The basic property of electrochemical current rectifier is that only unidirectional current can pass through, so the authors firstly immobilize some redox-active electron transfer mediators on the electrode surface with an insulating layer, generating weak current signal (Fig. 4.15). Then, the electron transfer mediators are modified, which can further modulate the electrochemical behavior of the solution-phase redox probes; thus, a unidirectional current signal from the solution-phase redox probes with much higher magnitude can be obtained. Furthermore, the current signal will be greatly decreased upon the binding of cells on the electrode, as the insulating cell membranes will block the electron transfer again. The response to HeLa cell concentration can be as low as 10 cell/mL, and the detection only requires 30 s.

The cancer cells treated with anticancer drugs have also been analyzed by using electrochemical technique. For instance, in the study conducted by El-Said et al., some HeLa cells are firstly fixed on gold film–covered silicon substrate.

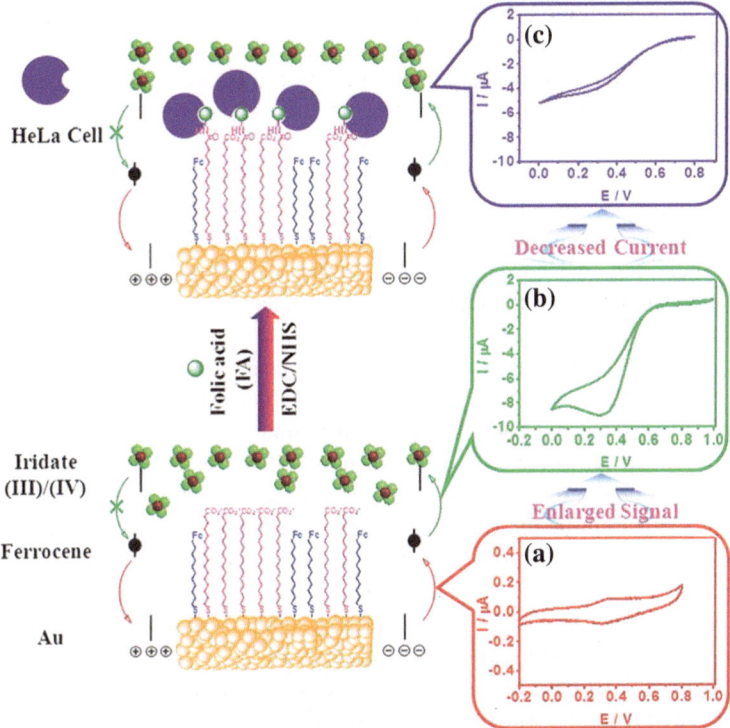

Fig. 4.15 Electron-transfer mechanism of cathodic electrochemical current rectification and the resultant signal amplification in the present system (from *a* to *b*). Folic acid was covalently immobilized to the electrode to target folate receptor–rich HeLa cells. A decreased signal was produced upon cell binding (from *b* to *c*). (Reproduced from Ref. [66] by permission of The Royal Society of Chemistry)

The immobilized HeLa cells may exhibit a good quasi-reversible cyclic voltammetric response, and the peak current increases with the number of the fixed cells. However, after the attached living cells are treated by anticancer drugs, the electrochemical response will be decreased, indicating the inhibition of cell growth [67]. Moreover, an increased concentration of the used anti-tumor drug will result in the decrease in the observed peak current.

4.5 Nanomaterials-Based Electrochemical Analysis of Cancer Cells

Due to the unique advantages of nanomaterials in optical, thermal, electrical, magnetic, mechanical and chemical properties, a growing number of nanomaterials have been employed for the electrochemical analysis of cancer cells, so we will specially summarize the achievements in this section, although some studies have been commented in the above sections.

Jia et al. [68] once proposed a method for fast magnetic separation, cell immobilization and electrochemical detection by using positively charged magnetic nanoparticles. While the magnetic nanoparticles attached to cancer cells could be isolated in a magnetic field and then tightly immobilized on an electrode surface, the endogenous electroactive species of the cells would transfer electron to the electrode, leading to an obvious anodic peak. They have also observed that growth of the cancer cell will be inhibited if an anticancer drug 5-fluorouracil is added in the test solution; thus, the endogenous substances will be reduced, making the peak current to be decreased.

Liu et al. [69] have reported an electrochemical method to detect the cancer cells rich in folate receptor by using folic acid–functionalized AuNPs. Since the folic acid–functionalized AuNPs have been immobilized on an electrode surface, due to the interaction between folic acid immobilized on AuNPs and its receptor overexpressed on tumor cell membrane, the cancer cells are then attached onto the electrode surface; thus, the electrochemical communication between the electroactive probes $K_3[Fe(CN)_6]/K_4[Fe(CN)_6]$ and the electrode will be changed. By making use of the electrochemical change, the folic acid–functionalized AuNPs–modified electrode can clearly denote folate receptor–positive tumor cells, such as ovarian tumor cells and human cervical cancer cells.

Zheng et al. [70] have also proposed an electrochemical method for the detection of folate receptor–positive tumor cells. As shown in Fig. 4.16, by making use of the specific recognition of the nanoprobes fabricated with polydopamine-coated CNTs and folate to cell surface folate receptor, the cancer cells can be electrochemically detected with electrochemical impedance spectroscopy technique. The proposed method can be used for the detection of as low as 500 cells, since the constructed nanoprobes may contribute a greatly enhanced electrochemical response.

Ding et al. [71] have fabricated a sandwich structure for the detection of K562 cells by making use of concanavalin globulin A (Con A)-modified electrode and AuNPs which are also modified with Con A. Since Con A modified on the surfaces of both the electrode and the nanoparticles can combine with the K562 cells, the sandwich structure can be constructed. The detection range of this method can be from 1.0×10^2 to 1.0×10^7 cells/mL. They have also used the fabricated sandwich structure to analyze polysaccharides on the K562 cell surface, and the amount of mannose moieties on a single living K562 cell is measured to correspond to 4.7×10^9 molecules of free mannose.

Chen et al. [72] have used MWCNT-modified electrode to analyze the change in electrochemical behavior of the K562 cancer cells treated by anti-tumor drugs. Since MWCNT can promote the electron transfer between the electroactive centers of the cells and the electrode, an irreversible voltammetric response at around +0.8 V can be observed in the first scan. Based on the obtained electrochemical wave, an electrochemical method is developed as an effective way to study the effect of anti-tumor drugs on cancer cells.

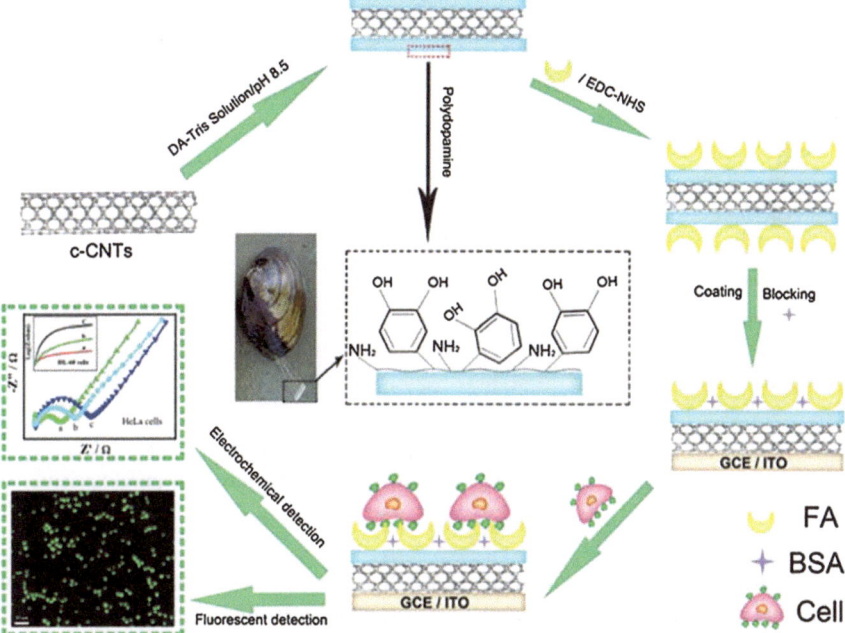

Fig. 4.16 Schematics of the folic acid–targeted cytosensing system. *FA* folic acid. (Reproduced from Ref. [70] by permission of The Royal Society of Chemistry)

Zhang et al. [73] have analyzed the expression of mannose and sialic acid by both normal cells and cancer cells with a lectin-based biosensor. They firstly fabricate the biosensor by preparing an AuNPs/MWCNT composite film on the surface of GCE. After lectins are further immobilized on the electrode surface, cells can be thus captured. Further studies reveal that the immobilized lectins can be quantitatively related to the cell surface glycosylation, and mannose may have a high expression level on the surface of both normal cells and cancer cells, while sialic acid shows a higher expression level only on cancer cell surface. Moreover, after the specific interaction between the surface-confined lectins and glycans on certain cell surface, lectin functionalized AuNPs are also introduced to form a sandwich-type structure. Due to the amplification effect of AuNPs, greatly enhanced electrochemical signals can be obtained, and the biosensor can be successfully utilized to quite sensitively quantify cancer cells and the average amount of sialic acid on a single cell.

Zhong et al. [74] have proposed another effective way to capture cells by using 3-aminophenylboronic acid–functionalized MWCNTs based on the sugar-specific affinitive interactions that result in the binding of K562 leukemia cells with functionalized MWCNTs via boronic acid groups. Since the CNTs may provide much rich binding sites for aminophenylboronic acid to interact with the cells, the detection sensitivity has been greatly improved. At the same time, efficient electronic conductivity of MWCNTs can make the electrochemical detection more conveniently operated.

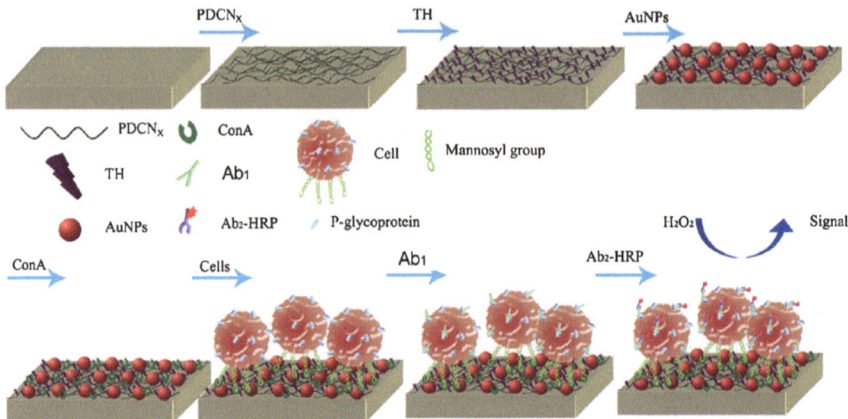

Fig. 4.17 Schematic illustration of the cell-based electrochemical enzyme-linked immunoassay. (Reprinted with the permission from Ref. [75]. Copyright 2010 American Chemical Society)

To sensitively analyze cell surface carbohydrate and glycoprotein, Zhang et al. [75] have designed an electrochemical cytosensor by using nitrogen-doped CNTs, thionine and AuNPs. As shown in Fig. 4.17, these materials are firstly immobilized on an electrode surface with layer-by-layer technique. Then, lectin Con A is adsorbed on the electrode surface, which is used as the first tier of signal amplification for the detection of HeLa cells. Since HeLa cells have numerous mannose residues, the cells can be combined with Con A; thus, the cancer cells are further immobilized on the electrode surface. Moreover, HRP-labeled p-glycoprotein antibody is introduced to bind with p-glycoprotein on the cell surface. Consequently, hydrogen peroxide can be catalytically oxidized by HRP to serve as the second tier of signal amplification. Based on the cascade signal amplification strategy, HeLa cells can be detected with a detection limit of 500 cells/mL and a liner range of $8.0 \times 10^2 \sim 2.0 \times 10^7$ cells/mL. In addition, with pre-sealing technique, the mannosyl groups and p-glycoprotein on single HeLa cell can be measured to be $(4 \pm 2) \times 10^{10}$ molecules of mannose moieties and 8.47×10^6 molecules of p-glycoprotein.

Cheng et al. [76] have also proposed an effective way to capture cancer cells with dual signal amplification for sensitive detection of cell surface carbohydrate. In this strategy, they make use of the specific recognition of the integrin receptors, which are over expressed on the surface of human gastric cancer cell line BGC-823, to arginine-glycine-aspartic acid-serine (RGDS) that is immobilized on single-walled CNTs (SWCNTs). As shown in Fig. 4.18, very fine electrochemical waves can be obtained at the HRP-Con A/BGC/RGDS-SWCNTs/GCE for different BGC cell concentrations. Based on the dual amplified signal by CNTs and enzymatic reaction catalyzed by HRP, this method can detect as low as 620 cells/ mL cancer cells. In addition, this research group has also generalized their strategies for dynamic analysis of carcinoma cell surface glycans [77].

Fig. 4.18 Differential pulse voltammograms of HRP-Con A/BGC/RGDS-SWCNTs/GCE obtained with BGC cell concentrations from 1×10^7 to 1×10^3 cells/mL (from *a* to *j*). Inset plot shows DPV peak current versus logarithm of BGC cell concentration. (Reprinted with the permission from Ref. [76]. Copyright 2008 American Chemical Society)

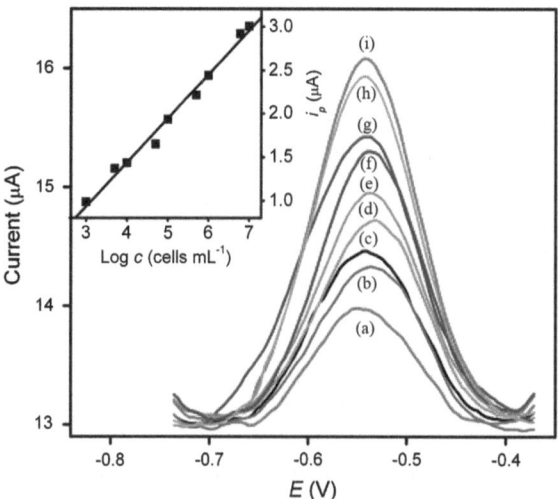

By using graphene and aptamer AS1411, Feng et al. [78] have developed a reusable and label-free electrochemical sensor for the detection of cancer cells. Since graphene may have excellent biocompatibility, which can maintain the biologic activity of cancer cells, and nucleolin overexpressed on the cancer cell surface can bind with AS1411 with high affinity, the detected cells can be fixed on the electrode surface, hindering the electron transfer between the electrode and the electrochemical probe $Fe(CN)_6^{3-/4-}$ in the solution. Based on the change in the electron transfer, detection of cancer cells can be achieved. Moreover, since a complementary nucleic acid may bind with the aptamer AS1411 more strongly, the fixed cancer cells can be detached from the aptamer; thus, the fabricated biosensor can be reused.

In living cells, carbohydrate molecules are involved in many important physiological processes, including specific tissue targeting, cell adhesion, cell differentiation, cell recognition of microbial pathogenicity and immune recognition [79]. Moreover, owing to their important role in the occurrence and migration of cancer, cell surface carbohydrates are often used as the target sites for tumor cell detection [80, 81]. Therefore, the electrochemical analysis of cell surface carbohydrates can not only help to understand their role in the development of diseases, but also contribute to the early diagnosis of tumor. Ding et al. [82] have utilized mannose to fabricate a sugar monolayer on an electrode surface. As shown in Fig. 4.19, based on the competitive binding with Con A between sugar on the electrode surface and polysaccharides on cancer cell surface, assay of cancer cell surface polysaccharide can be achieved. In this study, since leukemia cell line K562 rich in surface mannose can competitively combine with quantum dot–labeled Con A, the binding of Con A on electrode surface is reduced; thus, the electrochemical response is decreased accordingly. The proposed method to assay cell surface carbohydrate can be also used for the detection of cancer cells, the sensitivity of which can be as low as 10^2 cells/mL.

Fig. 4.19 Schematics of the assay of cell surface carbohydrate. *QD* quantum dot. Reprinted with the permission from Ref. [82]. Copyright 2008 American Chemical Society

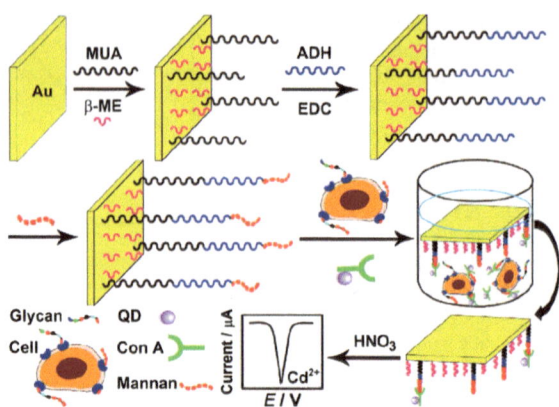

Combination of nanomaterials with aptamers has opened more opportunities for the electroanalysis of cells with high sensitivity. For instance, Li et al. [83] have used aptamer-quantum dots conjugates to design a competitive strategy for sensitive detection of cancer cells. In this strategy, the complementary DNA (cDNA) of the aptamer of the target cells is firstly immobilized on a gold electrode. Subsequently, aptamer-quantum dots conjugates are assembled by the formation of double-stranded DNA structure. Nevertheless, since the target cells may compete with the cDNA to bind with the aptamer-quantum dots conjugates, the quantum dots will be released from the electrode surface. Therefore, based on the measurement of the remaining aptamer-quantum dots, which can be determined by electrochemical stripping measurement of Cd^{2+} concentration in the quantum dots, the target cells can be detected with a very high sensitivity.

Ding et al. [84] have also combined nanomaterials with aptamers to design a strategy for electrochemical detection of cancer cells. They firstly fabricate some AuNPs bifunctionalized with aptamers and CdS nanoparticles. Then, they prepare some magnetic beads modified with the capture DNA molecules. Since the aptamer immobilized on AuNPs may hybridize with the capture DNA attached on magnetic beads, some complexes are formed. Nevertheless, in the presence of the detected cancer cells, AuNPs are released from the complexes, due to the high affinity between the aptamers immobilized on the AuNPs and their targets. Since the electrochemical signal can be amplified by the fabricated AuNPs bifunctionalized with aptamers and CdS nanoparticles, the detection limit can be 67 cells/mL, with a detection range from 1.0×10^2 to 1.0×10^5 cells/mL.

Combination of nanomaterials with aptamers may make the electrochemical analysis of cells to also have a very high selectivity. For instance, Li et al. [85] have proposed an electrochemical immunoassay to selectively detect breast cancer cells by simultaneously measuring two co-expressing tumor markers, human mucin-1 (MUC1) and CEA, on the surface of the cancer cells, by using the aptamer of MUC1 and CdS nanoparticles. As shown in Fig. 4.20, firstly, breast cells MCF-7 are attached on the electrode surface through the interaction between the aptamer previously immobilized on the electrode surface and the MUC1

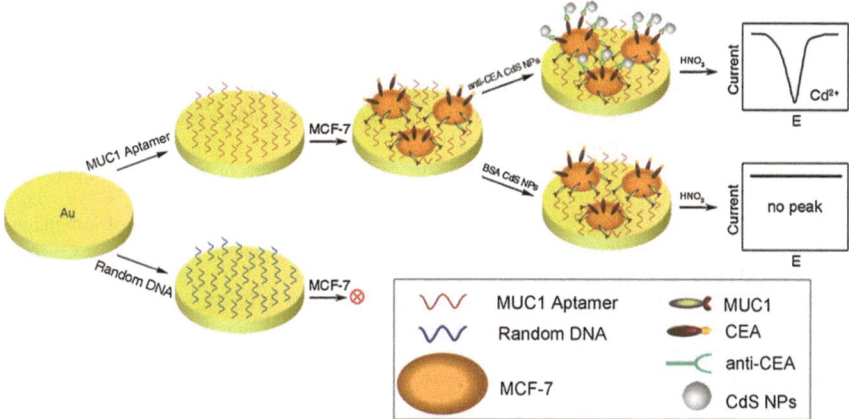

Fig. 4.20 Schematic representation of the method for the detection of breast cancer cells through simultaneous recognition of two different tumor markers. (Reprinted from Ref. [85], Copyright 2010, with permission from Elsevier)

protein expressed on the cell surface. Then, CdS nanoparticles–labeled CEA antibody recognizes and binds with the CEA protein, which is also expressed on the cancer cell surface. Subsequently, anodic stripping voltammetry is employed to detect cadmium stripping signal; thus, the cancer cells can be quantitatively detected. Since the cancer cells are analyzed by simultaneously detecting two tumor markers (MUC1 and CEA) on the cell surface, MCF-7 can be effectively detected without false-positive results. Meanwhile, with this method, breast cancer cell MCF-7 can be easily distinguished from other kinds of cells, such as acute leukemia cells CCRF-CEM and normal cells islet beta cells.

References

1. Huang JY, Zhang DM, Xing W, Ma X, Yin YX, Wei Q, Li GX (2008) An approach to assay calcineurin activity and the inhibitory effect of zinc ion. Anal Biochem 375(2):385–387
2. Miao P, Liu L, Li Y, Li GX (2009) A novel electrochemical method to detect mercury (II) ions. Electrochem Commun 11(10):1904–1907
3. Miao P, Liang ZQ, Liu L, Chen GF (2011) Fabrication of multi-functionalized gold nanoparticles and the application to electrochemical detection of nitrite. Curr Nanosci 7(3):354–358
4. Ronkainen NJ, Halsall HB, Heineman WR (2010) Electrochemical biosensors. Chem Soc Rev 39(5):1747–1763
5. Wang ZY, Liu L, Xu YY, Sun LZ, Li GX (2011) Simulation and assay of protein biotinylation with electrochemical technique. Biosens Bioelectron 26(11):4610–4613
6. Yang QL, Zhao J, Zhou ND, Ye ZH, Li GX (2011) Electrochemical sensing telomere-bending motions caused by hTRF1. Biosens Bioelectron 26(5):2228–2231
7. Zhao J, Meng WY, Miao P, Yu ZG, Li GX (2008) Photodynamic effect of hypericin on the conformation and catalytic activity of hemoglobin. Int J Mol Sci 9(2):145–153
8. Ding L, Du D, Zhang XJ, Ju HX (2008) Trends in cell-based electrochemical biosensors. Curr Med Chem 15(30):3160–3170

9. Popovtzer R, Neufeld T, Biran N, Ron EZ, Rishpon J, Shacham-Diamand Y (2005) Novel integrated electrochemical nano-biochip for toxicity detection in water. Nano Lett 5(6):1023–1027

10. Slaughter GE, Bieberich E, Wnek GE, Wynne KJ, Guiseppi-Elei A (2004) Improving neuron-to-electrode surface attachment via alkanethiol self-assembly: an alternating current impedance study. Langmuir 20(17):7189–7200

11. Nonner W, Eisenberg B (2000) Electrodiffusion in ionic channels of biological membranes. J Mol Liq 87(2–3):149–162

12. Huang JY, Chen L, Zhang X, Liu SL, Li GX (2008) Electrochemical studies of ion-channel behavior of annexin V in phosphatidylcholine bilayer membranes. Electrochem Commun 10(3):451–454

13. Alcantara K, Munge B, Pendon Z, Frank HA, Rusling JF (2006) Thin film voltammetry of spinach photosystem II. proton-gated electron transfer involving the Mn-4 cluster. J Am Chem Soc 128(46):14930–14937

14. Proux-Delrouyre V, Demaille C, Leibl W, Setif P, Bottin H, Bourdillon C (2003) Electrocatalytic investigation of light-induced electron transfer between cytochrome c(6) and photosystem I. J Am Chem Soc 125(45):13686–13692

15. Shao ML, Bai HJ, Gou HL, Xu JJ, Chen HY (2009) Cytosensing and evaluation of cell surface glycoprotein based on a biocompatible poly(diallydimethylammonium) doped poly(dimethylsiloxane) film. Langmuir 25(5):3089–3095

16. Du D, Ju HX, Zhang XJ, Chen J, Cai J, Chen HY (2005) Electrochemical immunoassay of membrane P-glycoprotein by immobilization of cells on gold nanoparticles modified on a methoxysilyl-terminated butyrylchitosan matrix. Biochemistry-Us 44(34):11539–11545

17. El-Ali J, Sorger PK, Jensen KF (2006) Cells on chips. Nature 442(7101):403–411

18. Meng FB, Yang JH, Liu T, Zhu XL, Li GX (2009) Electric communication between the inner part of a cell and an electrode: the way to look inside a cell. Anal Chem 81(21):9168–9171

19. Liu J, Zhou H, Xu JJ, Chen HY (2011) An effective DNA-based electrochemical switch for reagentless detection of living cells. Chem Commun 47(15):4388–4390

20. Balaban RS, Nemoto S, Finkel T (2005) Mitochondria, oxidants, and aging. Cell 120(4):483–495

21. Zhao J, Meng FB, Zhu XL, Han K, Liu SL, Li GX (2008) Electrochemistry of mitochondria: a new way to understand their structure and function. Electroanal 20(14):1593–1598

22. Zheng XT, Hu WH, Wang HX, Yang HB, Zhou W, Li CM (2011) Bifunctional electro-optical nanoprobe to real-time detect local biochemical processes in single cells. Biosens Bioelectron 26(11):4484–4490

23. Hartwell LH, Weinert TA (1989) Checkpoints: controls that ensure the order of cell-cycle events. Science 246(4930):629–634

24. Kafi MA, Kim TH, An JH, Choi JW (2011) Fabrication of cell chip for detection of cell cycle progression based on electrochemical method. Anal Chem 83(6):2104–2111

25. Pappas D, Martinez MM, Reif RD (2010) Detection of apoptosis: a review of conventional and novel techniques. Anal Methods-Uk 2(8):996–1004

26. Taylor RC, Cullen SP, Martin SJ (2008) Apoptosis: controlled demolition at the cellular level. Nat Rev Mol Cell Bio 9(3):231–241

27. Thatte U, Dahanukar S (1997) Apoptosis: clinical relevance and pharmacological manipulation. Drugs 54(4):511–532

28. Gerke V, Moss SE (2002) Annexins: from structure to function. Physiol Rev 82(2):331–371

29. Tong CY, Shi BX, Xiao XJ, Liao HD, Zheng YQ, Shen GL, Tang DY, Liu XM (2009) An Annexin V-based biosensor for quantitatively detecting early apoptotic cells. Biosens Bioelectron 24(6):1777–1782

30. Liu T, Zhu W, Yang X, Chen L, Yang RW, Hua ZC, Li GX (2009) Detection of apoptosis based on the interaction between annexin V and phosphatidylserine. Anal Chem 81(6):2410–2413

31. Li Z, Jo J, Jia JM, Lo SC, Whitcomb DJ, Jiao S, Cho K, Sheng M (2010) Caspase-3 activation via mitochondria is required for long-term depression and AMPA receptor internalization. Cell 141(5):859–871

32. Xiao H, Liu L, Meng FB, Huang JY, Li GX (2008) Electrochemical approach to detect apoptosis. Anal Chem 80(13):5272–5275
33. Zhang JJ, Zheng TT, Cheng FF, Zhu JJ (2011) Electrochemical sensing for caspase 3 activity and inhibition using quantum dot functionalized carbon nanotube labels. Chem Commun 47(4):1178–1180
34. Zhang JJ, Zheng TT, Cheng FF, Zhang JR, Zhu JJ (2011) Toward the early evaluation of therapeutic effects: an electrochemical platform for ultrasensitive detection of apoptotic cells. Anal Chem 83(20):7902–7909
35. Dubiel EA, Martin Y, Vermette P (2011) Bridging the gap between physicochemistry and interpretation prevalent in cell-surface interactions. Chem Rev 111(4):2900–2936
36. Gu HY, Chen Z, Sa RX, Yuan SS, Chen HY, Ding YT, Yu AM (2004) The immobilization of hepatocytes on 24 nm-sized gold colloid for enhanced hepatocytes proliferation. Biomaterials 25(17):3445–3451
37. Lee SJ, Choi JS, Park KS, Khang G, Lee YM, Lee HB (2004) Response of MG63 osteoblast-like cells onto polycarbonate membrane surfaces with different micropore sizes. Biomaterials 25(19):4699–4707
38. Choi CK, English AE, Jun SK, Kihm KD, Rack PD (2007) An endothelial cell compatible biosensor fabricated using optically thin indium tin oxide silicon nitride electrodes. Biosens Bioelectron 22(11):2585–2590
39. Xiao CD, Lachance B, Sunahara G, Luong JHT (2002) Assessment of cytotoxicity using electric cell-substrate impedance sensing: concentration and time response function approach. Anal Chem 74(22):5748–5753
40. Yang LJ, Li YB, Erf GF (2004) Interdigitated array microelectrode-based electrochemical impedance immunosensor for detection of Escherichia coli O157: H7. Anal Chem 76(4):1107–1113
41. Chen H, Heng CK, Puiu PD, Zhou XD, Lee AC, Lim TM, Tan SN (2005) Detection of Saccharomyces cerevisiae immobilized on self-assembled monolayer (SAM) of alkanethiolate using electrochemical impedance spectroscopy. Anal Chim Acta 554(1–2):52–59
42. Ding L, Hao C, Xue YD, Ju HX (2007) A bio-inspired support of gold nanoparticles-chitosan nanocomposites gel for immobilization and electrochemical study of K562 leukemia cells. Biomacromolecules 8(4):1341–1346
43. Ding L, Du D, Wu H, Ju HX (2007) A disposable impedance sensor for electrochemical study and monitoring of adhesion and proliferation of K562 leukaemia cells. Electrochem Commun 9(5):953–958
44. Hao C, Ding L, Zhang XJ, Ju HX (2007) Biocompatible conductive architecture of carbon nanofiber-doped chitosan prepared with controllable electrodeposition for cytosensing. Anal Chem 79(12):4442–4447
45. Maalouf R, Fournier-Wirth C, Coste J, Chebib H, Saikali Y, Vittori O, Errachid A, Cloarec JP, Martelet C, Jaffrezic-Renault N (2007) Label-free detection of bacteria by electrochemical impedance spectroscopy: comparison to surface plasmon resonance. Anal Chem 79(13):4879–4886
46. Ruan CM, Yang LJ, Li YB (2002) Immunobiosensor chips for detection of Escherichia coli O157: H7 using electrochemical impedance spectroscopy. Anal Chem 74(18):4814–4820
47. Varshney M, Li YB (2007) Interdigitated array microelectrode based impedance biosensor coupled with magnetic nanoparticle-antibody conjugates for detection of Escherichia coli O157: H7 in food samples. Biosens Bioelectron 22(11):2408–2414
48. Radke SM, Alocilja EC (2005) A microfabricated biosensor for detecting foodborne bioterrorism agents. IEEE Sens J 5(4):744–750
49. Du D, Cai J, Ju HX, Yan F, Chen J, Jiang XQ, Chen HY (2005) Construction of a biomimetic zwitterionic interface for monitoring cell proliferation and apoptosis. Langmuir 21(18):8394–8399
50. Cheng MS, Lau SH, Chow VT, Toh CS (2011) Membrane-based electrochemical nano-biosensor for escherichia coli detection and analysis of cells viability. Environ Sci Technol 45(15):6453–6459

51. Liu L, Miao P, Xu YY, Tian ZP, Zou ZG, Li GX (2010) Study of Pt/TiO(2) nanocomposite for cancer-cell treatment. J Photoch Photobio B 98(3):207–210

52. Yin YM, Cao Y, Xu YY, Li GX (2010) Colorimetric immunoassay for detection of tumor markers. Int J Mol Sci 11(12):5078–5095

53. Harley CB (2008) Telomerase and cancer therapeutics. Nat Rev Cancer 8(3):167–179

54. Yang QL, Nie YJ, Zhu XL, Liu XJ, Li GX (2009) Study on the electrocatalytic activity of human telomere G-quadruplex-hemin complex and its interaction with small molecular ligands. Electrochim Acta 55(1):276–280

55. Shao ZY, Liu YX, Xiao H, Li GX (2008) PCR-free electrochemical assay of telomerase activity. Electrochem Commun 10(10):1502–1504

56. Chen L, Huang JY, Meng FB, Zhou ND (2010) Distinguishing tumor cells via analyzing intracellular telomerase activity. Anal Sci 26(5):535–538

57. Yang WQ, Zhu X, Liu QD, Lin ZY, Qiu B, Chen GN (2011) Label-free detection of telomerase activity in HeLa cells using electrochemical impedance spectroscopy. Chem Commun 47(11):3129–3131

58. Li Y, Liu BW, Li X, Wei QL (2010) Highly sensitive electrochemical detection of human telomerase activity based on bio-barcode method. Biosens Bioelectron 25(11):2543–2547

59. Han K, Chen L, Lin ZS, Li GX (2009) Target induced dissociation (TID) strategy for the development of electrochemical aptamer-based biosensor. Electrochem Commun 11(1):157–160

60. Han K, Liang ZQ, Zhou ND (2010) Design strategies for aptamer-based biosensors. Sensors-Basel 10(5):4541–4557

61. Wang YL, Li D, Ren W, Liu ZJ, Dong SJ, Wang EK (2008) Ultrasensitive colorimetric detection of protein by aptamer: Au nanoparticles conjugates based on a dot-blot assay. Chem Commun 22:2520–2522

62. Ferreira CSM, Matthews CS, Missailidis S (2006) DNA aptamers that bind to MUC1 tumour marker: design and characterization of MUC1-binding single-stranded DNA aptamers. Tumor Biol 27(6):289–301

63. Shangguan D, Li Y, Tang ZW, Cao ZHC, Chen HW, Mallikaratchy P, Sefah K, Yang CYJ, Tan WH (2006) Aptamers evolved from live cells as effective molecular probes for cancer study. P Natl Acad Sci USA 103(32):11838–11843

64. Shao ZY, Li Y, Yang QL, Wang J, Li GX (2010) A novel electrochemical method to detect cell surface carbohydrates and target cells. Anal Bioanal Chem 398(7–8):2963–2967

65. de la Escosura-Muniz A, Sanchez-Espinel C, Diaz-Freitas B, Gonzalez-Fernandez A, Maltez-da Costa M, Merkoci A (2009) Rapid identification and quantification of tumor cells using an electrocatalytic method based on gold nanoparticles. Anal Chem 81(24):10268–10274

66. Li HL, Li D, Liu JY, Qin YN, Ren JT, Xu SL, Liu YQ, Mayer D, Wang EK (2012) Electrochemical current rectifier as a highly sensitive and selective cytosensor for cancer cell detection. Chem Commun 48(20):2594–2596

67. El-Said WA, Yea CH, Kim H, Oh BK, Choi JW (2009) Cell-based chip for the detection of anti-cancer effect on HeLa cells using cyclic voltammetry. Biosens Bioelectron 24(5):1259–1265

68. Jia XE, Tan L, Zhou YP, Jiang XF, Xie QJ, Tang H, Yao SZ (2009) Magnetic immobilization and electrochemical detection of leukemia K562 cells. Electrochem Commun 11(1):141–144

69. Liu L, Zhu XL, Zhang DM, Huang JY, Li GX (2007) An electrochemical method to detect folate receptor positive tumor cells. Electrochem Commun 9(10):2547–2550

70. Zheng TT, Zhang R, Zou LF, Zhu JJ (2012) A label-free cytosensor for the enhanced electrochemical detection of cancer cells using polydopamine-coated carbon nanotubes. Analyst 137(6):1316–1318

71. Ding CF, Qian SW, Wang ZF, Qu B (2011) Electrochemical cytosensor based on gold nanoparticles for the determination of carbohydrate on cell surface. Anal Biochem 414(1):84–87

72. Chen J, Du D, Yan F, Ju HM, Lian HZ (2005) Electrochemical antitumor drug sensitivity test for leukemia K562 cells at a carbon-nanotube-modified electrode. Chem-Eur J 11(5):1467–1472

73. Zhang XA, Teng YQ, Fu Y, Xu LL, Zhang SP, He B, Wang CG, Zhang W (2010) Lectin-based biosensor strategy for electrochemical assay of glycan expression on living cancer cells. Anal Chem 82(22):9455–9460

74. Zhong X, Bai HJ, Xu JJ, Chen HY, Zhu YH (2010) A reusable interface constructed by 3-aminophenylboronic acid functionalized multiwalled carbon nanotubes for cell capture, release, and cytosensing. Adv Funct Mater 20(6):992–999

75. Zhang JJ, Cheng FF, Zheng TT, Zhu JJ (2010) Design and implementation of electrochemical cytosensor for evaluation of cell surface carbohydrate and glycoprotein. Anal Chem 82(9):3547–3555

76. Cheng W, Ding L, Lei JP, Ding SJ, Ju HX (2008) Effective cell capture with tetrapeptide-functionalized carbon nanotubes and dual signal amplification for cytosensing and evaluation of cell surface carbohydrate. Anal Chem 80(10):3867–3872

77. Cheng W, Ding L, Ding SJ, Yin YB, Ju HX (2009) A simple electrochemical cytosensor array for dynamic analysis of carcinoma cell surface glycans. Angew Chem Int Edit 48(35):6465–6468

78. Feng LY, Chen Y, Ren JS, Qu XG (2011) A graphene functionalized electrochemical aptasensor for selective label-free detection of cancer cells. Biomaterials 32(11):2930–2937

79. Ohtsubo K, Marth JD (2006) Glycosylation in cellular mechanisms of health and disease. Cell 126(5):855–867

80. Dai Z, Kawde AN, Xiang Y, La Belle JT, Gerlach J, Bhavanandan VP, Joshi L, Wang J (2006) Nanoparticle-based sensing of glycan-lectin interactions. J Am Chem Soc 128(31):10018–10019

81. Meany DL, Hackler L, Zhang H, Chan DW (2011) Tyramide signal amplification for antibody-overlay lectin microarray: a strategy to improve the sensitivity of targeted glycan profiling. J Proteome Res 10(3):1425–1431

82. Ding L, Cheng W, Wang XJ, Ding SJ, Ju HX (2008) Carbohydrate monolayer strategy for electrochemical assay of cell surface carbohydrate. J Am Chem Soc 130(23):7224–7225

83. Li JJ, Xu M, Huang HP, Zhou JJ, Abdel-Halim ES, Zhang JR, Zhu JJ (2011) Aptamer-quantum dots conjugates-based ultrasensitive competitive electrochemical cytosensor for the detection of tumor cell. Talanta 85(4):2113–2120

84. Ding CF, Ge Y, Zhang SS (2010) Electrochemical and electrochemiluminescence determination of cancer cells based on aptamers and magnetic beads. Chem-Eur J 16(35):10707–10714

85. Li T, Fan Q, Liu T, Zhu XL, Zhao J, Li GX (2010) Detection of breast cancer cells specially and accurately by an electrochemical method. Biosens Bioelectron 25(12):2686–2689